LONDON MATHEMATICAL SOCIETY STUDENT TEXTS

Managing Editor: Professor D. Benson,
Department of Mathematics, University of Aberdeen, UK

London Mathematical Society Student Texts 80

The Riemann Hypothesis for Function Fields

Frobenius Flow and Shift Operators

MACHIEL VAN FRANKENHUIJSEN

Utah Valley University

CAMBRIDGE
UNIVERSITY PRESS

CAMBRIDGE
UNIVERSITY PRESS

University Printing House, Cambridge CB2 8BS, United Kingdom

One Liberty Plaza, 20th Floor, New York, NY 10006, USA

477 Williamstown Road, Port Melbourne, VIC 3207, Australia

314-321, 3rd Floor, Plot 3, Splendor Forum, Jasola District Centre, New Delhi - 110025, India

103 Penang Road, #05-06/07, Visioncrest Commercial, Singapore 238467

Cambridge University Press is part of the University of Cambridge.

It furthers the University's mission by disseminating knowledge in the pursuit of education, learning and research at the highest international levels of excellence.

www.cambridge.org
Information on this title: www.cambridge.org/9781107685314

First published 2014

A catalogue record for this publication is available from the British Library

ISBN 978-1-107-04721-1 Hardback
ISBN 978-1-107-68531-4 Paperback

To my beautiful wife Jena

Contents

Illustrations

Preface

This book grew out of an attempt to understand the paper [Conn1], in which Alain Connes constructs a beautiful noncommutative space with a view to proving the Riemann hypothesis. That paper is supplemented by Shai Haran's papers [Har2, Har3], which give a similar construction with more details on some of the computations. Connes' proof is explored in Chapter 6, where his method is applied with an aim of proving the Riemann hypothesis for a curve over a finite field (Weil's theorem).

Chapter 5 presents Bombieri's proof [Bom1] of the Riemann hypothesis for curves over a finite field. This chapter is not necessary for Chapter 6, and can be skipped by a reader who is only interested in understanding Connes' approach.

Chapters 1, 2, and 3 provide background. Chapter 1 is an exposition of the theory of valued fields, and in Chapters 2 and 3, we present Tate's thesis [Ta] for curves over a finite field.

There are numerous exercises throughout the book where the reader is asked to work out a detail or explore related material. The exercises that are labelled as 'problems' ask questions that may not have a definite answer.

This book is not primarily about number fields, but occasionally we discuss the connection between number fields and function fields. We have included several diagrams to help the reader create a mental picture of this connection.

The author believes that Connes' approach provides the first truly convincing heuristic argument for the Riemann hypothesis. He also believes that working out this argument for the function field case is the key to getting it to work for the integers. It is therefore not surprising that we do not reach our goal in Chapter 6. This book provides the basis for further research in this direction.

Acknowledgements The research for this book was started at the University of California in Riverside and continued over the years at Rutgers University (New Jersey), Utah Valley State College (now Utah Valley University), the Institut des Hautes Études Scientifiques (IHÉS) in Bures-sur-Yvette, France, of which I was a member during February of 2007, and the Georg-August-Universität in Göttingen, Germany, during the author's sabbatical year at the invitation of Professor Dr. Ralf Meyer in 2011.

The material and financial support of the IHÉS and the Department of Mathematics and the School of Science and Health of Utah Valley University is gratefully ackowledged. The generous support from Professor Meyer during the author's sabbatical year is also gratefully acknowledged. I also want to acknowledge the many excellent teachers whom I had at the university of Nijmegen, Riverside, the IHÉS, Rutgers University, and Göttingen. I fondly remember Serge Lang, whose way of doing mathematics has been an example ever since we first met. I miss him. I want to thank the students and professors who gave talks in my seminar in Göttingen when I was there in 2011. That seminar greatly accelerated the evolution of this book.

Introduction

The Riemann zeta function is the function $\zeta(s)$, defined for $\operatorname{Re} s > 1$ by

$$\zeta(s) = 1 + \frac{1}{2^s} + \frac{1}{3^s} + \cdots .$$

It has a meromorphic continuation to the complex plane, with a simple pole at $s = 1$ with residue 1. Completing this function with the factor for the archimedean valuation, $\zeta_{\mathbb{R}}(s) = \pi^{-s/2}\Gamma(s/2)$, one obtains the function

$$\zeta_{\mathbb{Z}}(s) = \zeta_{\mathbb{R}}(s)\zeta(s). \tag{1}$$

The function $\zeta_{\mathbb{Z}}$ is meromorphic on $\mathbb{C}$ with simple poles at 0 and 1, and it satisfies the functional equation $\zeta_{\mathbb{Z}}(1 - s) = \zeta_{\mathbb{Z}}(s)$. This is proved by Riemann [Ri1] using the "Riemann–Roch" formula[1]

$$\theta(t^{-1}) = t\theta(t), \tag{2}$$

where $\theta(t) = \sum_{n=-\infty}^{\infty} e^{-\pi n^2 t^2}$ is closely related to the so-called theta-function.[2] For $\operatorname{Re} s > 1$, the zeta function satisfies the Euler product

$$\zeta_{\mathbb{Z}}(s) = \zeta_{\mathbb{R}}(s) \prod_p \frac{1}{1 - p^{-s}}, \tag{3}$$

where the product is taken over all prime numbers. It follows that the zeros of $\zeta_{\mathbb{Z}}$ all lie in the vertical strip $0 \leq \operatorname{Re} s \leq 1$.[3] The Riemann

---

[1] Formula (2) goes back to Cauchy and is called "Riemann–Roch Theorem" in Tate's thesis.
[2] The classical theta function is defined as $\vartheta(z, \tau) = \sum_{n=-\infty}^{\infty} e^{\pi i n^2 \tau} e^{2\pi i n z}$. Its relation to θ is $\theta(t) = \vartheta(0, it^2)$.
[3] It is also known that the zeros do not lie on the boundary of this "critical strip" [In, Theorem 19, p. 58] and [vF1, Theorem 2.4].

hypothesis states that these zeros all lie on the line $\mathrm{Re}\, s = 1/2$:

$$\text{Riemann hypothesis:} \quad \zeta_{\mathbb{Z}}(s) = 0 \text{ implies } \mathrm{Re}\, s = 1/2.$$

See [Ri1] and [Ed, Har2, Lap-vF1, Lap-vF2, Ta] for more information about the Riemann and other zeta functions.

In this exposition, we prove the Riemann hypothesis for the zeta function of a nonsingular curve over a finite field. Let q be a power of a prime number p, and let $m(T, X)$ be a polynomial in two variables with coefficients in $\mathbb{F}_q$, the finite field with q elements. The equation $m(T, X) = 0$ defines a curve $\mathcal{C}$ over $\mathbb{F}_q$, which we assume to be nonsingular. Let $N_\mathcal{C}(n)$ be the number of solutions of the equation $m(t, x) = 0$ in the finite set $\mathbb{F}_{q^n} \times \mathbb{F}_{q^n}$. Thus $N_\mathcal{C}(n)$ is the number of points on $\mathcal{C}$ with coordinates in $\mathbb{F}_{q^n}$. A famous theorem of F. K. Schmidt in 1931 (see [fSch] and [Has1, Has2, Tr]) says that there exist an integer g, the genus of $\mathcal{C}$, and algebraic numbers $\omega_1, \ldots, \omega_{2g}$, such that[4]

$$N_\mathcal{C}(n) = q^n - \sum_{\nu=1}^{2g} \omega_\nu^n + 1.$$

We also define $N_\mathcal{C}(0) = 2 - 2g$. From the formula for $N_\mathcal{C}(n)$, the Mellin transform (generating function)

$$\mathcal{M}N_\mathcal{C}(s) = \sum_{n=0}^{\infty} N_\mathcal{C}(n) q^{-ns}$$

can be computed as a rational function of q^{-s},

$$\mathcal{M}N_\mathcal{C}(s) = \frac{1}{1 - q^{1-s}} - \sum_{\nu=1}^{2g} \frac{1}{1 - \omega_\nu q^{-s}} + \frac{1}{1 - q^{-s}}.$$

We define the *zeta function of $\mathcal{C}$* by

$$\zeta_\mathcal{C}(s) = q^{s(g-1)} \frac{\prod_{\nu=1}^{2g}(1 - \omega_\nu q^{-s})}{(1 - q^{1-s})(1 - q^{-s})},$$

so that $\mathcal{M}N_\mathcal{C}$ is recovered as its logarithmic derivative,

$$-\frac{\zeta_\mathcal{C}'(s)}{\zeta_\mathcal{C}(s)} = \big(\mathcal{M}N_\mathcal{C}(s) + g - 1\big)\log q.$$

---

[4] In $N_\mathcal{C}(n)$, also the finitely many points "at infinity" need to be counted. See Chapter 5.

The function $\zeta_{\mathcal{C}}$ satisfies the functional equation $\zeta_{\mathcal{C}}(1-s) = \zeta_{\mathcal{C}}(s)$. This functional equation can be proved using the Riemann–Roch theorem

$$l(D) = \deg D + 1 - g + l(\mathcal{K} - D), \tag{4}$$

which is the analogue of (2) above. This theorem is proved in Chapter 3.

Since the zeta function of $\mathcal{C}$ is a rational function of q^{-s}, it is periodic with period $2\pi i / \log q$. It has two simple poles at $s = 0$ and 1, and $2g$ zeros at $s = \log_q \omega_\nu$, all repeated modulo $2\pi i / \log q$. It satisfies an Euler product, analogous to (3), which converges for $\operatorname{Re} s > 1$,

$$\zeta_{\mathcal{C}}(s) = q^{s(g-1)} \prod_v \frac{1}{1 - q^{-s \deg v}},$$

where the product is taken over all orbits of the Frobenius flow on $\mathcal{C}$, and $\deg v$ is the length of an orbit. It follows that $1 \leq |\omega_\nu| \leq q$. Artin conjectured[5] that $\zeta_{\mathcal{C}}(s)$ has its zeros on the line $\operatorname{Re} s = 1/2$. In terms of the exponentials of the zeros, the numbers ω_ν, this means that

$$|\omega_\nu| = \sqrt{q} \ \text{ for every } \nu = 1, \ldots, 2g.$$

This is the analogue of the Riemann hypothesis for $\zeta_{\mathcal{C}}$. It is trivially verified for $\mathcal{C} = \mathbb{P}^1$, when $g = 0$ and $\zeta_{\mathcal{C}}$ does not have any zeros. It was proved by H. Hasse in the case of elliptic curves ($g = 1$), and first in full generality by A. Weil.[6] Later proofs, based on one of Weil's proofs, have been given by P. Roquette [Roq] and others. Weil uses the intersection of divisors with the graph of Frobenius in $\mathcal{C} \times \mathcal{C}$, and his second proof uses the action of Frobenius on the embedding of $\mathcal{C}$ into its Jacobian (see [Ros, Appendix]). There have been some attempts to translate the first and second proof to the Riemann zeta function, when $\mathcal{C}$ is the "curve" $\operatorname{spec} \mathbb{Z}$, but so far without success, one of the obstacles being that in the category of schemes, $\operatorname{spec} \mathbb{Z} \times \operatorname{spec} \mathbb{Z}$ is one-dimensional and not two-dimensional as the surface $\mathcal{C} \times \mathcal{C}$ (see [Har1]).

A completely new technique was discovered by Stepanov [Ste], initially only for hyperelliptic curves. W. M. Schmidt [wSch] used his method to reprove the Riemann hypothesis for $\zeta_{\mathcal{C}}$, and a simplified proof was given by Bombieri [Bom1,Bom2]. Bombieri's proof uses the graph of Frobenius

---

[5] In his thesis [Art1], Artin considers only quadratic extensions of $\mathbb{F}_p(T)$, that is, hyperelliptic curves over $\mathbb{F}_p$. Moreover, in his zeta functions, the Euler factors corresponding to the points at infinity are missing. Later, F. K. Schmidt introduced the zeta function of a general projective curve over an arbitrary finite field [fSch].

[6] Weil announced his ideas in 1940 [W1,W2] and explained them in 1942 in a letter to Artin [W3]. But the complete proof (see [W5, W7]) had to await the completion of his *Foundations* [W4]. See also [Ray].

in $\mathcal{C} \times \mathcal{C}$.[7] It relies on the Riemann–Roch formula and also on the action of Frobenius. So his proof uses less geometry that cannot be translated to spec $\mathbb{Z}$. We present this proof in Chapter 5.

Around the same time, P. Deligne proved the Weil conjectures [Del, FreiK, K2]. His proof, applied in the one-dimensional case, gives yet another proof of the Riemann hypothesis for curves over finite fields. This proof relies on detailed information about the action of Frobenius on the étale cohomology groups of the variety (see [K1, K3] for more information). As with the other proofs, it is unclear how to translate this proof to the number field case.

Recently, Alain Connes found a completely new method again, based on the work of Shai Haran [Har1] (and then extended by Haran in the papers [Har2, Har3]), using harmonic analysis on the ring of adeles. Connes establishes the positivity of the trace of a certain shift operator, thus proving the Riemann hypothesis for spec $\mathbb{Z}$ (and for all L-functions associated with Grössencharacters), up to a "lemma" about a noncommutative space. In Chapter 6, we adapt Haran's approach to obtain a proof of the Riemann hypothesis for $\zeta_{\mathcal{C}}$. This proof does not use $\mathcal{C} \times \mathcal{C}$ or Frobenius. Indeed, the only difficulty in translating it to a proof for spec $\mathbb{Z}$ is to provide a suitable local analysis at the real component of the adeles. Connes does this by using special functions on $[-c, c]$, the Fourier transforms of which are also almost supported on $[-c, c]$. In our case, working with the curve $\mathcal{C}$, we have no archimedean valuations to deal with, and the local analysis at the primes "at infinity" is not different from the local analysis at the other points of $\mathcal{C}$.

It is interesting to see the development in these proofs. Gradually, more geometry that cannot be translated to the number field case has been taken out. The question arises what exactly is needed to prove the Riemann hypothesis for curves, and what can we learn from this for the Riemann hypothesis for spec $\mathbb{Z}$.

Weil's first proof uses the geometry of $\mathcal{C} \times \mathcal{C}$, and, in particular, the intersection of the graph of Frobenius with the diagonal. There is some reason to believe that no analogue will ever exist for number fields, or, at least, that constructing an analogue is harder than establishing the Riemann hypothesis. His second proof uses the Jacobian of $\mathcal{C}$, and, again, no direct analogue may ever be constructed for the integers.

Deligne's proof works for higher-dimensional varieties and uses very detailed information about the geometry, along with sophisticated tools

---

[7] Bombieri does not mention $\mathcal{C} \times \mathcal{C}$, but he uses $\mathbb{F}_q^a(\mathcal{C}) \otimes \mathbb{F}_q^a(\mathcal{C})$, which is the ring of functions on $\mathcal{C} \times \mathcal{C}$.

to make deductions from this information. It is not impossible that an adequate analogue of these cohomological theories exists for the integers (see [Den1, Den2]).

Bombieri's proof uses very little of the geometry of $\mathcal{C} \times \mathcal{C}$, but it uses the action of Frobenius and Riemann–Roch. Since Tate's thesis, it is known that the Riemann–Roch equality translates into formula (2). Bombieri's proof naturally divides into two steps. In the first step, he uses the action of Frobenius to obtain a discrete flow on the curve, which is analyzed to obtain an upper bound for the number of points on the curve, which is a weak form of the prime number theorem for the curve (see Table 5.1 in Section 5.5). One sees that the horizontal coordinate of $\mathcal{C} \times \mathcal{C}$ plays an "arithmetic" role, and the vertical coordinate plays a "geometric" role (see Remark 5.4.7). In the second step, the Riemann hypothesis for $\mathbb{P}^1$ is used, along with the fact that the Frobenius automorphism generates the local Galois group (decomposition group) at a point on the curve, to obtain a lower bound for the number of points on the curve. Combining the two steps, one first obtains the analogue of the prime number theorem with a good error term, and from this it is a small step to deduce the Riemann hypothesis. Therefore, one might conclude that the right approach to the Riemann hypothesis is to first prove the prime number theorem with a good bound for the error.

Looking at the first step of Bombieri's proof, one might even guess that the key to a prime number theorem with a good error bound is to construct a function (possibly a Fourier or Dirichlet polynomial, in the spirit of the methods of Baker, Gelfond, and Schneider) that vanishes at the first N primes to a high order. If one could bound the degree of this polynomial, then one would obtain an upper bound for the number of primes. This would already imply the Riemann hypothesis, so the second step becomes unnecessary. According to Deninger [Den1, Den2], the analogue of the Frobenius flow of the first step might be provided by the shift on the real line.[8]

Connes, in turn, does not use the geometry of $\mathcal{C} \times \mathcal{C}$ or Frobenius. Instead, he uses Fourier analysis on the ring of adeles and the diagonal embedding of the global field (the field of rational numbers). Even though he does not use the Riemann–Roch theorem, this theorem is a

---

[8] According to Bombieri, the correct philosophy is as follows: since Frobenius in characteristic p is based on the fact that the binomial coefficients $\binom{p}{k}$ are divisible by p if $k \neq 0, p$, an understanding of the archimedean Frobenius should come from an understanding of the size of $\binom{n}{k}$. This leads to the Gaussian $e^{-\pi x^2}$ and probabilities [Bom2] (personal communication, October 2008).

natural consequence of the formalism that he sets up, much as in Tate's thesis. The only problem is that on the real line, the Fourier transform of a compactly supported function is not compactly supported.[9] To complete the proof of the Riemann hypothesis for spec $\mathbb{Z}$, one might try to construct a suitable substitute for this requirement.

The fact that Connes does not use the action of Frobenius should make one suspicious about his approach. However, Connes uses the action of the idele class group on the space of adele classes. As Connes points out [Conn1, Remark c, p. 72], this action is the counterpart of the action of Frobenius on the curve. Indeed, by class field theory, there exists an isomorphism between the exact sequences

$$
\begin{array}{ccccc}
\ker & \longrightarrow & \mathbb{A}^*/K^* & \xrightarrow{\;|\cdot|\;} & q^{\mathbb{Z}} \\
\| & & \| & & \| \\
\pi_{\mathrm{ab}}(\mathcal{C}) & \longrightarrow & G_{\mathrm{ab}} & \longrightarrow & \langle \phi \rangle
\end{array}
\tag{5}
$$

The vertical isomorphisms come from class field theory. The upper sequence gives the norm from the idele class group $\mathbb{A}^*/K^*$ to the group of powers of q. The kernel of this map corresponds to the group $\pi_{\mathrm{ab}}(\mathcal{C})$ of abelian covers of the curve $\mathcal{C}$ inside the Galois group of abelian extensions

$$
G_{\mathrm{ab}} = \mathrm{Gal}(\mathbb{F}_q(\mathcal{C})^{\mathrm{ab}}, \mathbb{F}_q(\mathcal{C})).
$$

On the level of Galois theory (the lower sequence in (5)), the second arrow maps G_{ab} onto the group generated by the Frobenius automorphism ϕ. This automorphism acts on constant field extensions of the function field of $\mathcal{C}$, which corresponds to the Frobenius flow on $\mathcal{C}$.

The counterpart for $\mathbb{Z}$ is the diagram of exact sequences

$$
\begin{array}{ccccc}
\mathbb{A}^*/\mathbb{Q}^*\mathbb{R}^+ & \longrightarrow & \mathbb{A}^*/\mathbb{Q}^* & \xrightarrow{\;|\cdot|\;} & \mathbb{R}^+ \\
\| & & \| & & \| \\
\mathrm{Gal}(\mathbb{Q}^{\mathrm{ab}}, \mathbb{Q}) & \longrightarrow & ? & \longrightarrow & ?
\end{array}
\tag{6}
$$

where the left vertical map is the isomorphism from class field theory between the group $\mathbb{A}^*/\mathbb{Q}^*\mathbb{R}^+$ and the Galois group of the maximal abelian extension of $\mathbb{Q}$. For the other vertical equalities, no counterpart is known

---

[9] In the field of p-adic numbers, and in general in any nonarchimedean field, the Fourier transform of a function that is locally constant is compactly supported. This is not true for the archimedean fields $\mathbb{R}$ and $\mathbb{C}$. See Section 3.1.

within Galois theory. Indeed, such a counterpart must lie outside of Galois theory, since any continuous map $\mathbb{R}^+ \to \mathrm{Gal}(\mathbb{Q}^a/\mathbb{Q})$ is constant, the latter group being profinite. Thus, only the "geometric part" of diagram (5), the abelian fundamental group $\pi_{\mathrm{ab}}(\mathcal{C})$, has its counterpart in number theory, while the "arithmetic part," corresponding to constant field extensions, remains a mystery in number theory.[10] However, in [Conn2], Connes defines the noncommutative space $\mathbb{A}/\mathbb{Q}^*$ of the adeles modulo the multiplicative action of the diagonally embedded rationals. This space has a natural multiplicative action of the idele class group by shift operators, thus completing the question marks in (6). In particular, the question mark under $\mathbb{R}^+$ corresponds to the multiplicative action of $\mathbb{R}^+$ on $\mathbb{A}/\mathbb{Q}^*$ by shifts, which, by diagram (5), corresponds to the action of Frobenius. It seems that everything is in place to prove the Riemann hypothesis for spec $\mathbb{Z}$.

This seems to be the first time since the formulation of the Riemann hypothesis in 1859 that we have a serious heuristic argument for its truth. It is surprising that such an easily stated problem, either as "all nonreal zeros of $\zeta(s)$ have real part $1/2$" or as "the prime number theorem has an error term of order $x^{1/2+\varepsilon}$," should be so hard to solve. Indeed, this was not immediately appreciated at the time, since Barnes assigned the Riemann hypothesis to Littlewood as a thesis problem (see [Conr]). But from the solution of the Riemann hypothesis for curves over finite fields, it seems that there may never be a proof using only methods from analytic number theory.[11]

---

[10] Every extension of $\mathbb{Q}$ (abelian or not) is ramified at some primes, hence should be considered as a geometric cover of curves. On the other hand, every abelian extension of $\mathbb{Q}$ is cyclotomic, hence could be considered as a constant field extension. It seems that the first two Galois groups in (5) have collapsed into the one group $\mathrm{Gal}(\mathbb{Q}^{\mathrm{ab}}, \mathbb{Q})$, and the group $\langle \phi \rangle$ is missing.

[11] For a possible approach using the theory of fractal strings, see [vF1, Remark 4.5].

1
Valuations

Valuations correspond to orbits of points on a curve: every orbit of Frobenius gives a valuation, and every valuation gives an orbit. This will be elaborated in Chapter 5. In this chapter, we study how valuations distinguish constants from nonconstant functions and how valuations can ramify.

We first develop this theory for an arbitrary extension L/K. In Section 1.4, we apply our theory to the situation that is the subject of this book, where L/K is a finite extension $K/\mathbf{q}$ of the field of rational functions $\mathbf{q} = \mathbb{F}_q(T)$ over the finite field of constants $\mathbb{F}_q$.

The embedding $\mathbf{q} \hookrightarrow K$ corresponds to the projection $\mathcal{C} \twoheadrightarrow \mathbb{P}^1$, where a point on the curve $\mathcal{C}$ projects to the value of the function T at this point. This will be explained in Chapter 5.

1.1 Trace and norm

Let L/K be a finite extension of fields. We can find a polynomial

$$m(X) = X^n + m_1 X^{n-1} + \cdots + m_n,$$

irreducible over K, such that $L = K[X]/(m)$.

For $x \in L$, we consider the map $M_x \colon L \to L$ of multiplication by x,

$$M_x(y) = xy \qquad (x, y \in L).$$

This map is linear in y and consequently has a determinant and a trace, defined independently of a choice of a basis for L as a K-vector space.

Definition 1.1.1   Let L/K be a finite separable extension, and $x \in L$. The *trace* of x over K is the trace of M_x,

$$\mathrm{Tr}_{L/K}(x) = \mathrm{Tr}(M_x).$$

The *norm* of x over K is the determinant of M_x, $N_{L/K}(x) = \det(M_x)$.

In particular, let $x = X + (m)$ be the root of m in L. The matrix of M_x on the basis $(1, x, \ldots, x^{n-1})$ of L over K is given by

$$M_x = \begin{pmatrix} 0 & & O & -m_n \\ 1 & \ddots & & \vdots \\ & \ddots & 0 & -m_2 \\ O & & 1 & -m_1 \end{pmatrix}.$$

The characteristic polynomial of this matrix is $m(X)$. Assume that L/K is separable, i.e., all roots of m are different. Then, over a splitting field of m, the matrix M_x diagonalizes. Denoting by $\sigma_1 x, \ldots, \sigma_n x$ the images of x in a splitting field F under the n embeddings $\sigma_1, \ldots, \sigma_n$ of L into F, we find that this matrix diagonalizes as

$$M_x = \begin{pmatrix} \sigma_1 x & & O \\ & \ddots & \\ O & & \sigma_n x \end{pmatrix} \qquad (x = X + (m)), \qquad (1.1)$$

on a suitable basis for F^n.

Exercise 1.1.2   Show that $\det(\lambda I - M_x) = m(\lambda)$.

Exercise 1.1.3 L is n-dimensional over K, and F^n is n-dimensional over F. Show that $K[X]/(m) \otimes F \cong F[X]/(m) \cong F \oplus \cdots \oplus F$.

Clearly, for any two elements $x, y \in L$, we have $M_{x+y} = M_x + M_y$ and $M_{xy} = M_x M_y$. Hence $M_{f(x)} = f(M_x)$ for every polynomial f over K. An element $y \in L$ can be written as $y = f(X) + (m)$ for a polynomial f over K. Therefore, $M_y = f(M_x)$ for M_x as in (1.1), and the matrix of M_y diagonalizes as

$$M_y = \begin{pmatrix} f(\sigma_1 x) & & O \\ & \ddots & \\ O & & f(\sigma_n x) \end{pmatrix} = \begin{pmatrix} \sigma_1 y & & O \\ & \ddots & \\ O & & \sigma_n y \end{pmatrix},$$

since $f(\sigma_i x) = \sigma_i f(x) = \sigma_i y$. We deduce the following proposition and corollary:

Proposition 1.1.4 *Let L/K be a finite separable extension of fields, and let K^a be an algebraic closure of K. Then*

$$\mathrm{Tr}_{L/K}(x) = \sum_{\sigma:L\to K^a} \sigma x,$$

where the summation extends over all embeddings of L into K^a.

For a finite separable extension M of L, each embedding $\sigma: L \to K^a$ extends to an embedding of M into K^a in $[M:L]$ different ways. Thus we obtain the following corollary:

Corollary 1.1.5 $\mathrm{Tr}_{M/K} = \mathrm{Tr}_{L/K} \circ \mathrm{Tr}_{M/L}$ *in a tower $M/L/K$ of finite separable extensions.*

This property of the trace will be needed in Chapter 3 to define the additive character of a field.

Exercise 1.1.6 Formulate and prove the analogue of Proposition 1.1.4 and Corollary 1.1.5 for the norm.

1.1.1 The canonical pairing

We have a K-valued pairing between elements of L, given by

$$(x, y) \longmapsto \mathrm{Tr}_{L/K}(xy).$$

Writing Tr for $\mathrm{Tr}_{L/K}$, the matrix of this pairing on a basis $(\epsilon_1, \ldots, \epsilon_n)$ for L as a K-vector space is

$$D_\epsilon = \begin{pmatrix} \mathrm{Tr}(\epsilon_1^2) & \mathrm{Tr}(\epsilon_1\epsilon_2) & \cdots & \mathrm{Tr}(\epsilon_1\epsilon_n) \\ \mathrm{Tr}(\epsilon_2\epsilon_1) & \mathrm{Tr}(\epsilon_2^2) & \cdots & \mathrm{Tr}(\epsilon_2\epsilon_n) \\ \vdots & \vdots & \ddots & \vdots \\ \mathrm{Tr}(\epsilon_n\epsilon_1) & \mathrm{Tr}(\epsilon_n\epsilon_2) & \cdots & \mathrm{Tr}(\epsilon_n^2) \end{pmatrix}.$$

The pairing of two elements x and y can be computed using D_ϵ as follows. Write $x = x_1\epsilon_1 + \cdots + x_n\epsilon_n$ and $y = y_1\epsilon_1 + \cdots + y_n\epsilon_n$ with coefficients x_i and y_i in K. Then

$$\mathrm{Tr}_{L/K}(xy) = (x_1 \ \ldots \ x_n)\, D_\epsilon \begin{pmatrix} y_1 \\ \vdots \\ y_n \end{pmatrix}.$$

By Proposition 1.1.4, $D_\epsilon = S_\epsilon^t S_\epsilon$, where S_ϵ is the matrix

$$S_\epsilon = \begin{pmatrix} \sigma_1 \epsilon_1 & \cdots & \sigma_1 \epsilon_n \\ \vdots & \ddots & \vdots \\ \sigma_n \epsilon_1 & \cdots & \sigma_n \epsilon_n \end{pmatrix}, \tag{1.2}$$

and S_ϵ^t denotes its transpose. On the standard basis $(1, x, x^2, \ldots, x^{n-1})$ with $x = X + (m)$, we find that $S_x = (\sigma_i x^{j-1})_{i,j}$. Hence $\det S_x$ is the Vandermonde determinant $\prod_{k=1}^n \prod_{l>k} (\sigma_l x - \sigma_k x)$. Since $\sigma_l x \neq \sigma_k x$ for $l \neq k$ by separability, we conclude that the canonical pairing is nondegenerate.

Theorem 1.1.7   *The image of* $\mathrm{Tr}_{L/K}$ *is* K.

Proof   Since the canonical pairing is nondegenerate, $t = \mathrm{Tr}_{L/K}(l)$ does not vanish for some $l \in L$. Moreover, it follows directly from the definition that $\mathrm{Tr}_{L/K}$ is a K-linear map, hence we find that $\mathrm{Tr}_{L/K}(xl/t) = x$ for every $x \in K$.   □

1.2 Valued fields

A *norm* on a field K is a function $|\cdot|: K \to [0, \infty)$ such that for some constant $C \geq 1$ the *triangle inequality*

$$|x + y| \leq C \max\{|x|, |y|\} \tag{1.3}$$

is satisfied, and further $|x| = 0$ if and only if $x = 0$, and $|xy| = |x| \cdot |y|$ for every $x, y \in K$.

The corresponding *valuation* on K is defined by

$$v(x) = -\log |x|.$$

The norm is called *nonarchimedean* if the triangle inequality holds with $C = 1$, and it is called *archimedean* if $|2| > 1$ (see also Lemma 1.3.1). In the second case, we can replace the norm by a power such that $|2| = 2$. Then we can take $C = 2$ in (1.3) and show that

$$|x + y| \leq |x| + |y|. \tag{1.4}$$

In terms of the valuation, the triangle inequality reads

$$v(x + y) \geq -c + \min\{v(x), v(y)\} \tag{1.5}$$

for some constant $c \geq 0$. If this inequality holds with $c = 0$, then v is nonarchimedean. If we replace v by a positive multiple of v, then (1.5) still holds, replacing c by the same multiple of c.

Exercise 1.2.1 Show that if v is nonarchimedean and $v(x) \neq v(y)$, then $v(x + y) = \min\{v(x), v(y)\}$. Thus every triangle is isosceles.

Remark 1.2.2 A nonarchimedean valuation corresponds to the order of vanishing of the function x at the "point" v, and we will usually normalize a valuation so that the group of values $\{v(x) : x \in K^*\}$ equals $\mathbb{Z}$.

A large value of $v(x)$ means that x vanishes at v to a high order, so that x is to be considered as small "around v." The norm of x is then close to 0. On the other hand, a large negative value for $v(x)$ means that x has a pole of high order at v. Then x is large "around v," and $|x| \gg 1$.

1.2.1 Norms on a vector space

A norm on a finite-dimensional vector space V over the normed field K is a non-negative valued function $\|\cdot\|$ that satisfies the triangle inequality

$$\|x + y\| \leq \|x\| + \|y\| \quad \text{for every } x, y \in V, \tag{1.6}$$

and further $\|x\| = 0$ if and only if $x = 0$, and $\|\lambda x\| = |\lambda| \cdot \|x\|$ for every $\lambda \in K$ and $x \in V$.

Remark 1.2.3 In this chapter, we are only interested in the topology induced by the norm on a vector space, so the normalization of the norm does not play an important role. Therefore, we simply think of $|\lambda|$ as the distance from λ to 0 in K, so that $\|x\|$ is the distance from x to $\vec{0}$ in V. In Chapter 3, however, it will be most natural to choose the norm on K so that $|\lambda|$ equals the factor by which the volume changes under multiplication by λ. This makes a difference only for the norm on $\mathbb{C}$, where we let $|\lambda| = |a + bi| = \sqrt{a^2 + b^2}$ in this chapter, which is the length of λ, whereas, in Chapter 3, we let $|\lambda| = a^2 + b^2$, which is proportional to the volume of the disc with radius $\sqrt{a^2 + b^2}$. The last choice of norm does not satisfy the triangle inequality (1.6), but it does satisfy (1.3) with $C = 4$.

Remark 1.2.4 We assume that (1.6) is satisfied, as opposed to the analogue of (1.3), so that it is easier to see that the balls around points of $x \in V$ form the basis of a topology on V. By [Art2, Chapters 1 and 2], this is no restriction. The following problem asks you to verify this.

Problem 1.2.5 Show that the condition $\|x + y\| \leq 2 \max\{\|x\|, \|y\|\}$ is equivalent to (1.6).

Let $B_r(x) = \{y : \|x - y\| < r\}$ denote the open ball of radius r and center x. The balls $B_r(x)$ form a basis for a topology on V for which addition and scalar multiplication are continuous. We assume that the closure of every ball in V is compact.

Lemma 1.2.6 *A set $B \subset V$ has compact closure if and only if it is bounded in norm.*

Proof Suppose $\bar{B}$ is compact. Since $\bar{B} \subseteq \bigcup_{n \in \mathbb{N}} B_n(0)$ is an open cover, we find that $\bar{B} \subseteq B_n(0)$ for some n. Hence $\|x\| < n$ for every $x \in B$.

Suppose now that the set B is bounded in norm, say $B \subseteq B_r(0)$. By the assumption of local compactness, the closure of $B_r(0)$ is compact. Hence B has compact closure.   □

Another norm $\| \cdot \|_1$ is equivalent to $\| \cdot \|$ if there exist positive constants C and C' such that

$$C'\|x\|_1 \le \|x\| \le C\|x\|_1$$

for every $x \in V$.

Lemma 1.2.7 *Let K be a locally compact complete valued field and let V be a K-vector space of finite dimension. Then V is locally compact and any two norms on V are equivalent.*

Proof Denote the norm on K by $|\cdot|$. As a finite-dimensional K-vector space, V is isomorphic to K^n for some n. Let $(e_1, \ldots, e_n)$ be a basis for V, so that a vector $(x_1, \ldots, x_n) \in K^n$ corresponds to $x_1 e_1 + \cdots + x_n e_n \in V$. We thus obtain the supremum norm on V,

$$\|x_1 e_1 + \cdots + x_n e_n\|_\infty = \max\{|x_i| : 1 \le i \le n\}.$$

By [Sa, Theorem 2.5], V with this norm is locally compact.

Let $\| \cdot \|$ be another norm on V. We obtain an inequality between $\| \cdot \|$ and $\| \cdot \|_\infty$ using the triangle inequality (1.6). Let $C = \|e_1\| + \cdots + \|e_n\|$. Then

$$\|x_1 e_1 + \cdots + x_n e_n\| \le |x_1| \cdot \|e_1\| + \cdots + |x_n| \cdot \|e_n\|$$
$$\le C\|x_1 e_1 + \cdots + x_n e_n\|_\infty.$$

Conversely, if there is no constant C' such that $C'\|x\|_\infty \le \|x\|$ for every $x \in V$, then we can choose a sequence $x^{(n)}$ of vectors of unit supremum norm such that $\|x^{(n)}\| \le 1/n$. By local compactness in the supremum norm, we can assume, after choosing a subsequence if necessary, that $x^{(n)}$ converges to a vector x, of unit supremum norm. In particular, $x \neq 0$. Then, for every n, $x = x^{(n)} + \left(x - x^{(n)}\right)$. Taking norms in V, we find, using the first part,

$$\|x\| \leq \left\|x^{(n)}\right\| + \left\|x - x^{(n)}\right\| \leq 1/n + C\left\|x - x^{(n)}\right\|_{\infty}.$$

Since $x^{(n)} \to x$ in the supremum norm, this implies that $\|x\| = 0$, and hence $x = 0$. This contradiction shows that $C'\|\cdot\|_{\infty} \leq \|\cdot\|$ for some $C' > 0$. Hence every norm on V is equivalent to $\|\cdot\|_{\infty}$. $\square$

In terms of valuations on a field, two valuations v and w are equivalent if and only if the function $x \mapsto |v(x) - w(x)|$ is bounded. More generally, v and w give the same topology on K if there exists a positive constant c such that $|v - cw|$ is bounded. For fields, the notion of equivalence reduces to a positive multiple, by the following lemma.

Lemma 1.2.8 *Let v and w be valuations of K. Then the topologies on K induced by v and w coincide if and only if w is a positive multiple of v.*

Proof Suppose $v(x) < 0$ and $w(x) \geq 0$. Then $x^n \to \infty$ in the topology of v, and x^n is bounded in the topology of w. Hence, if the topologies are the same, then v and w have the same sign on every element of K.

Let now $x \in K^*$ be such that $v(x) > 0$, and define the necessarily positive number c by

$$w(x) = cv(x).$$

Suppose that $w(y) < cv(y)$ for some $y \in K$. Replacing y by a large power, we may even assume that

$$w(y) < cv(y) - cv(x).$$

Let n be such that $0 \leq v(y) + nv(x) < v(x)$. Then $v(yx^n) \geq 0$. On the other hand, $w(yx^n) < cv(y) - cv(x) + ncv(x) < 0$. Hence v and w have opposite signs at yx^n. This contradiction shows that $w(y) \geq cv(y)$. But similarly, $w(y) > cv(y)$ is also impossible. We conclude that $w = cv$. $\square$

Let L/K be an extension of fields. We say that a valuation w on L extends the valuation v of K, notation $w \mid v$, if w restricted to K is equivalent to v. By the previous lemma, this means that there exists a positive real number c such that $w(x) = cv(x)$ for every $x \in K$.

In Section 1.3, we discuss the valuations of $\mathbb{F}_q(T)$. By the next lemma, this describes the valuations on every function field over a finite field.

Lemma 1.2.9 *Let L be a finite separable algebraic extension of a locally compact field K, and let v be a valuation of K. Then there exists* *a valuation of L extending v. This valuation is unique up to a positive multiple.*

Proof By Lemma 1.2.7, the topology on L as a vector space over K is unique, hence by Lemma 1.2.8, the extension of v to L is unique up to a positive multiple. To prove the existence of an extension, we define a function w on L by

$$w(x) = v(N_{L/K}(x)),$$

where $N_{L/K}$ denotes the norm of Definition 1.1.1. Then clearly $w(x) = \infty$ if and only if $x = 0$, and $w(xy) = w(x) + w(y)$. Also, w restricts to a multiple of v, since for $x \in K$, $w(x) = [L : K]v(x)$. To show the triangle inequality (1.5), it suffices to show that $w(1 + x)$ is bounded on the set of x with $w(x) \geq 0$. Since the set of such x is compact and addition is continuous, the set $\{1 + x \colon w(x) \geq 0\}$ is compact, and hence bounded in norm by Lemma 1.2.6. We conclude that w is a valuation on L extending v.   □

For $x \in K^a$ let L be a finite field extension containing x (for example, take $L = K[x]$). By Exercise 1.1.6, $v(N_{L/K}(x))/[L : K]$ is independent of the choice of L. Defining

$$v^a(x) = v(N_{L/K}(x))/[L : K],$$

we obtain the following corollary:

Corollary 1.2.10 *Let K^a be an algebraic closure of the locally compact field K. Then there is a unique valuation v^a on K^a extending v.*

Note that K^a may be infinite-dimensional over K, in which case it is not locally compact.

1.2.2 Discrete valuations

Let v be a nonarchimedean valuation of the locally compact field K. By the triangle inequality (1.5) (with $c = 0$) the set

$$\mathfrak{o}_v = \{x \in K \colon v(x) \geq 0\}$$

is a ring. We call it the ring of integers of K, or, since elements of K are interpreted as functions, the ring of functions that are *regular* at the point v. The units of this ring are the elements with $v(x) = 0$, and the ring has a unique maximal ideal of functions vanishing at v (that is, $v(x) > 0$).

We assume that the value group $\{v(x)\colon x \neq 0\}$ is a discrete subgroup of $\mathbb{R}$. It follows that there exists a function π_v, called a *uniformizer*, that vanishes at v to minimal order. Then the maximal ideal is $\pi_v \mathfrak{o}_v$.

The value at v of a regular function x is the element $x + \pi_v \mathfrak{o}_v$ of the *residue class field*

$$K(v) = \mathfrak{o}_v / \pi_v \mathfrak{o}_v.$$

It follows from the local compactness of K that $K(v)$ is a finite field.

Let L be a finite algebraic extension of K. By Lemma 1.2.9, there is a unique valuation w on L extending v. The finite field $L(w)$ is an extension of $K(v)$, of degree

$$f(w/v) = [L(w) : K(v)],$$

called the *degree of inertia*. We write π_w for a uniformizer of L. The *order of ramification* is the positive integer $e(w/v)$ such that

$$\pi_v \mathfrak{o}_w = \pi_w^{e(w/v)} \mathfrak{o}_w. \tag{1.7}$$

We say that L is *unramified* over K if $e(w/v) = 1$. Then π_v is also a uniformizer for w.

Exercise 1.2.11 Show that $e(w/v) = w(\pi_v)/w(\pi_w)$. We usually normalize w such that $w(\pi_w) = 1$, and then $e(w/v) = w(\pi_v)$.

Lemma 1.2.12 *The degree of L over K is $e(w/v)f(w/v)$.*

Proof Write $e = e(w/v)$ and $f = f(w/v)$. Choose representatives $\epsilon_1, \ldots, \epsilon_f$ in $\mathfrak{o}_w$ so that the classes in $L(w)$ form a basis over $K(v)$. Clearly, the system of ef numbers

$$\epsilon_i \pi_w^{j-1} \qquad (i = 1, \ldots, f, \ j = 1, \ldots, e) \tag{1.8}$$

is independent over K. Moreover, given an element $x \in L$, we first multiply it by a power of π_v so that $\pi_v^n x \in \mathfrak{o}_w$. Then we approximate $\pi_v^n x$ first modulo π_w by a combination over K of the numbers ϵ_i. Subtracting this approximation and dividing by π_w, we can compute the next digit of this approximation. We compute in this way e successively closer approximations to $\pi_v^n x$ modulo π_w^j $(j = 1, \ldots, e)$. The difference of $\pi_v^n x$ and the last approximation is divisible by π_v. Dividing by π_v, we continue, to obtain a sequence of linear combinations of the numbers (1.8), the coefficients of which converge in K. Hence the system is complete. $\square$

Remark 1.2.13 The system (1.8) provides a basis for $\mathfrak{o}_w$ as an $\mathfrak{o}_v$-module.

Lemma 1.2.14 *For a finite extension L/K of complete local fields with valuations $w \mid v$,*

$$\mathrm{Tr}_{L/K}(\mathfrak{o}_w) \subseteq \mathfrak{o}_v \quad and \quad \mathrm{Tr}_{L/K}(\pi_w \mathfrak{o}_w) \subseteq \pi_v \mathfrak{o}_v,$$

and every embedding $\sigma \colon L \to K^a$ that is the identity on K is continuous.

Proof Extend w to a valuation of K^a. By the uniqueness of the topology of L and Lemma 1.2.8, we see that $w(\sigma x) = w(x)$ for every embedding σ of L in K^a. Hence σ is continuous, and the rest of the lemma follows by Proposition 1.1.4. $\qquad\square$

1.2.3 Different and ramification

Given a ring $\mathfrak{o}$ with field of fractions K, a fractional ideal is a subset of K that is finitely generated as an $\mathfrak{o}$-module. Recall that for a fractional ideal, the *inverse* of $\mathfrak{a}$ is the fractional ideal defined by

$$\mathfrak{a}^{-1} = \{x \colon x\mathfrak{a} \subseteq \mathfrak{o}\}.$$

Exercise 1.2.15 Show that the inverse of $\mathfrak{a}^{-1}$ is $\mathfrak{a}$.

We define the inverse different (or *codifferent* in [Se]) first as a fractional ideal, and the different is then its inverse. The different turns out to be an ideal of $\mathfrak{o}_w$.

Definition 1.2.16 The *inverse different* of w is

$$\mathfrak{d}_{w/v}^{-1} = \{x \in L \colon \mathrm{Tr}_{L/K}(xy) \in \mathfrak{o}_v \text{ for every } y \in \mathfrak{o}_w\}.$$

By Lemma 1.2.14, $\mathfrak{o}_w \subseteq \mathfrak{d}_{w/v}^{-1}$. Clearly also, the inverse different is an $\mathfrak{o}_w$-module, and by Theorem 1.1.7, it is not all of L. Hence there exists an integer $d(w/v) \geq 0$, the *differential exponent*, such that

$$\mathfrak{d}_{w/v} = \pi_w^{d(w/v)} \mathfrak{o}_w. \tag{1.9}$$

We say that the different is *trivial* if $\mathfrak{d}_{w/v} = \mathfrak{o}_w$. Equivalently, the different is trivial if $d(w/v) = 0$.

In a trivial extension, the different is trivial: $d(v/v) = 0$ for every valuation v.

Theorem 1.2.17 *Let L/K be an extension of valued fields with valuations $w \mid v$. Then w is unramified over K if and only if the different is trivial.*

We prove this by relating both the different and ramification to the canonical pairing of Section 1.1.1 (see also [Art2, Chapter 5]).

By Lemma 1.2.14, if $\epsilon_1, \ldots, \epsilon_n$ is a basis of $\mathfrak{o}_w$ over $\mathfrak{o}_v$ (such bases exist by Remark 1.2.13), then the matrix of the canonical pairing has entries in $\mathfrak{o}_v$. We choose such a basis of L over K in the following two lemmas.

Lemma 1.2.18   *The valuation w of L is unramified over K if and only if D_ϵ is invertible modulo π_v.*

Proof   Suppose that w is ramified. Choosing the special basis (1.8), and by Lemma 1.2.14, we see that D_ϵ has rows that have a factor π_v. Hence D_ϵ is not invertible modulo π_v. On the other hand, if w is unramified, then the matrix of the canonical pairing of the extension of finite fields $L(w)/K(v)$ is given by D_ϵ modulo π_v. Since an extension of finite fields is separable, this matrix is nonsingular, so that D_ϵ is invertible modulo π_v.   □

Lemma 1.2.19   *The different is trivial if and only if D_ϵ is invertible modulo π_v. Moreover, if $\mathfrak{o}_v\epsilon_1 + \cdots + \mathfrak{o}_v\epsilon_n \subseteq \mathfrak{o}_w$ and D_ϵ is invertible modulo π_v, then $\mathfrak{o}_v\epsilon_1 + \cdots + \mathfrak{o}_v\epsilon_n = \mathfrak{o}_w$.*

Proof   Suppose that $\epsilon_1, \ldots, \epsilon_n$ is a basis of L over K consisting of elements of $\mathfrak{o}_w$, and suppose that D_ϵ is invertible modulo π_v. Let $x \in \mathfrak{d}_{w/v}^{-1}$. We can find elements $x_i \in K$ and an integer m such that $\min_i v(x_i) = 0$ and $x = \pi_v^m(x_1\epsilon_1 + \cdots + x_n\epsilon_n)$. Since $(x_i)^t D_\epsilon$ is not the zero vector modulo π_v, we can find an element $y = y_1\epsilon_1 + \cdots + y_n\epsilon_n$ with $y_i \in \mathfrak{o}_v$ so that $(x_i)^t D_\epsilon(y_i) = 1$. Then we have $\mathrm{Tr}_{L/K}(xy) = \pi_v^m$. Since this lies in $\mathfrak{o}_v$ by the assumption that $x \in \mathfrak{d}_{w/v}^{-1}$, it follows that $m \geq 0$. Hence x is an element of $\mathfrak{o}_v\epsilon_1 + \cdots + \mathfrak{o}_v\epsilon_n$. We find that

$$\mathfrak{o}_v\epsilon_1 + \cdots + \mathfrak{o}_v\epsilon_n \subseteq \mathfrak{o}_w \subseteq \mathfrak{d}_{w/v}^{-1} \subseteq \mathfrak{o}_v\epsilon_1 + \cdots + \mathfrak{o}_v\epsilon_n.$$

This proves the last statement and that the different is trivial if D_ϵ is invertible.

Conversely, let $\epsilon_1, \ldots, \epsilon_n$ be an $\mathfrak{o}_v$-basis of $\mathfrak{o}_w$, and suppose that D_ϵ is not invertible modulo π_v. Then we can choose a vector (x_i) in the kernel of D_ϵ modulo π_v such that $\min_i v(x_i) = 0$. Put $x = x_1\epsilon_1 + \cdots + x_n\epsilon_n$. Since $(x_i)^t D_\epsilon$ is a multiple of π_v, we have that $\mathrm{Tr}_{L/K}(xy) \in \pi_v\mathfrak{o}_v$ for every $y \in \mathfrak{o}_w$. By linearity of the trace, we obtain that $\mathrm{Tr}_{L/K}((x/\pi_v)y) \in \mathfrak{o}_v$. Hence $x/\pi_v \in \mathfrak{d}_{w/v}^{-1}$. Since $x/\pi_v \notin \mathfrak{o}_w$, it follows that the different is not trivial.   □

Combining these two lemmas, Theorem 1.2.17 follows.

1.2.4 Inseparable extensions

We define the derivative on $K[X]$ formally by $(X^n)' = nX^{n-1}$, extended by linearity over K.

Exercise 1.2.20   Prove that a polynomial $m(X)$ has a multiple root if and only if $m'(X)$ has the same root.

Suppose that the polynomial $m(X) \in K[X]$ is irreducible and has multiple roots. Then m' has lower degree than m and has a root in common with m. Since m is irreducible, it follows that $m' = 0$. By linearity, if aX^n is a monomial in m then m' contains the monomial anX^{n-1}. Since m' is the sum of such monomials, $m' = 0$ if and only if $n = 0$ in K for each monomial in m.

Definition 1.2.21   Let L/K be a finite extension of fields. An element of L is *inseparable over* K if its defining equation has multiple roots. The extension L/K is *inseparable* if L has inseparable elements. It is *purely inseparable* if every element of L lies in K or is inseparable over K.

By the foregoing discussion, in characteristic zero, $(X^n)' = 0$ only for the monomial X^0, hence every element is separable.

In characteristic p, $m' = 0$ implies that m contains only monomials of the form aX^{pn}. Then m can be written as $m(X) = f(X^p)$ for some polynomial f. If f is still inseparable, we repeat this process to obtain $m(X) = f(X^{p^s})$ for some $s \geq 1$.

Lemma 1.2.22   *Let L/K be a finite extension of fields of characteristic p, and let $x \in L$ be such that $x \notin K$ and $x^p \in K$. Then $K^{1/p}$, defined as $K[\sqrt[p]{K}]$, is a subfield of L of degree p over K, and $K^{1/p} = K[x]$.*

Proof   If $K^{1/p}$ has degree more than p over K, then we can find $y \in K^{1/p}$ such that $[K[y] : K] > p$. Since in characteristic p, taking the p-th root is a linear function, and y is a linear combination of p-th roots, it is itself a p-th root. Hence $y^p \in K$. This contradiction shows that $K^{1/p}$ is of degree at most p over K. Since $K[x]$ is a subfield of $K^{1/p}$, the lemma follows.   □

Exercise 1.2.23   Show that the subfield of L generated by all $x \in L$ such that x is separable over K is a separable extension L^{sep} of K. Show that L/L^{sep} is purely inseparable: every element x of L satisfies an irreducible equation $x^{p^s} = l$ for some $l \in L^{\mathrm{sep}}$.

By this exercise and the lemma, we obtain that $L = \left(L^{\mathrm{sep}}\right)^{1/p^s}$.

Definition 1.2.24 The field L^{sep} is called the *separable closure* of K in L.

Theorem 1.2.25 *Let v be a valuation of K and let L/K be a purely inseparable extension of degree p^s. Then there is a unique extension of v to L. The order of ramification of this extension is p^s, and the residue class field extension is trivial.*

Proof For each $x \in L$, x^{p^s} is an element of K. Hence, if w is an extension of v, then

$$w(x) = p^{-s}v\big(x^{p^s}\big).$$

It follows that the extension of v is unique.

Let π be a uniformizer for v. By the above formula, the value group of w is

$$p^{-s}v(\pi)\mathbb{Z}.$$

Since $\pi^{1/p^s} \in L$ and $w\big(\pi^{1/p^s}\big) = p^{-s}v(\pi)$, this element is a uniformizer for w. It follows that $e(w/v) = p^s$. By Lemma 1.2.12, $f(w/v) = 1$. $\square$

Remark 1.2.26 We refer to the monograph [V, Section 5.2] for a thorough account of the theory of inseparable extensions.

1.3 Valuations of $\mathbb{F}_q(T)$

A field contains 1, and therefore $n = 1 + \cdots + 1$ (n summands) and its negative $-n$. Thus every field contains an image of $\mathbb{Z}$. However, the map $\mathbb{Z} \to K$ is not always an embedding. For example, if $K = \mathbb{F}_q(T)$, then the kernel of this map is $p\mathbb{Z}$, where p is the prime factor of q. This is one example where the condition of the next lemma is satisfied, since the image of $\mathbb{Z}$ is finite.

Lemma 1.3.1 *Let v be a valuation of the field K. If v is bounded from below on the image of $\mathbb{Z}$ in K then v is nonarchimedean.*

Proof Let $v(m) \geq -B$ for every $m \in \mathbb{Z}$. The expression

$$(x+y)^{2^n-1} = \sum_{k=0}^{2^n-1} \binom{2^n-1}{k} x^k y^{2^n-1-k}$$

has 2^n terms. We apply the triangle inequality (1.5) n times, each time splitting the sum into two sums of equal length. We obtain

$$(2^n-1)v(x+y) \geq -nc - B + (2^n-1)\min\{v(x), v(y)\}.$$

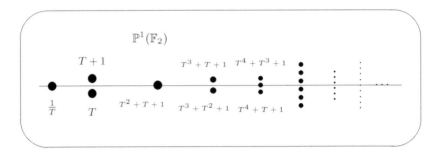

Figure 1.1 $\mathbb{P}^1(\mathbb{F}_2)$. The point at infinity corresponds to $\frac{1}{T}$.

Dividing by $(2^n - 1)$ and taking the limit for n to infinity, we obtain the nonarchimedean triangle inequality $v(x + y) \geq \min\{v(x), v(y)\}$. $\square$

The simplest function field is $\mathbb{F}_q(T)$, the extension of $\mathbb{F}_q$ generated by one transcendental element. Because of the analogy with $\mathbb{Q}$, we write

$$\mathbb{F}_q(T) = \mathbf{q}.$$

We define the degree of a quotient $f(T) = \alpha(T)/\beta(T)$ of two polynomials α and β by $\deg(f) = \deg(\alpha) - \deg(\beta)$. One valuation of $\mathbf{q}$ is the order of vanishing at infinity, the negative of the degree,

$$v_\infty(f) = -\deg(f),$$

In the proof of the next theorem, this valuation provides our first insight towards the construction of the so-called P-adic valuations: for each irreducible polynomial P, we can write a nonzero rational function as

$$f(T) = P(T)^m \frac{\alpha(T)}{\beta(T)},$$

where α and β are nonzero polynomials without a factor P. Then

$$v_P(f) = m \qquad\qquad\qquad (1.10)$$

defines the P-adic valuation of f.

See Figure 1.1 for a depiction on the projective line of the valuations of K. Each irreducible polynomial corresponds to a valuation and to $\deg P$ points on $\mathbb{P}^1(\mathbb{F}_q)$, and $1/T$ corresponds to the point at ∞ on this line. See Remark 1.3.4.

Theorem 1.3.2 *Let v be a valuation of $\mathbb{F}_q(T)$. Then v is trivial, or a positive multiple of v_∞, or of v_P for some irreducible polynomial P.*

Proof Let $d = d(T)$ be a polynomial of degree at least one (i.e., d is not constant). Let $f(T) = f_n(T)d^n + \cdots + f_0(T)$ be any polynomial written in base d, where the digits f_i are polynomials of degree $\deg f_i < \deg d$, and $f_n \neq 0$. Then $\deg f \geq n \deg d$. Since, by the previous lemma, the valuation v is nonarchimedean, we find that

$$v(f) \geq C + n \min\{0, v(d)\},$$

where $C = \min\{v(f) \colon f \in \mathbb{F}_q[T], \deg f < \deg d\}$, the minimum over a finite set. We obtain that $v(f) \geq C + \frac{\deg f}{\deg d} \min\{0, v(d)\}$. Applying this to f^m and letting $m \to \infty$, we obtain

$$v(f) \geq \frac{\deg(f)}{\deg(d)} \min\{0, v(d)\} \tag{1.11}$$

for every polynomial f.

First suppose that $v(d) \geq 0$ for some nonconstant polynomial d. By (1.11), we obtain that $v(f) \geq 0$ for every polynomial f. If $v(f) = 0$ for every nonzero polynomial, then $v(f) = 0$ for every nonzero rational function, hence v is the trivial valuation.

Otherwise, if v is not trivial, let then P be a polynomial of minimal degree such that $v(P) > 0$. If $P = \alpha\beta$ is a factorization into polynomials, then $v(\alpha)$ or $v(\beta)$ must be positive. By the minimality of P, we find that either α or β must be of the same degree as P, and the other factor is then a constant. Hence P is irreducible. Let now α be any polynomial that is not divisible by P. Dividing by P, we can find polynomials q and r such that $\alpha = qP + r$, where the remainder r is nonzero and $\deg r < \deg P$. Again by the minimality of P, it follows that $v(r) = 0$. Since $v(qP) > v(q) \geq 0$, we obtain by Exercise 1.2.1 that $v(\alpha) = 0$. For a general rational function $f = P^m \alpha/\beta$, where the polynomials α and β are not divisible by P, we see that $v(f) = mv(P)$. Therefore, v is a positive multiple of v_P, where P is the irreducible polynomial (unique up to a constant in $\mathbb{F}_q$) such that $v(P) > 0$.

Suppose now that $v(d) < 0$ for some, and hence for every polynomial of positive degree. Then we obtain from (1.11) that

$$\frac{v(f)}{\deg f} \geq \frac{v(d)}{\deg d}$$

for every nonconstant polynomial f. If f is not constant, then we can also write d in the base f to obtain the opposite inequality. Therefore $v(f)/\deg f$ is a constant, independent of f. It follows that v is a positive multiple of the valuation v_∞. $\square$

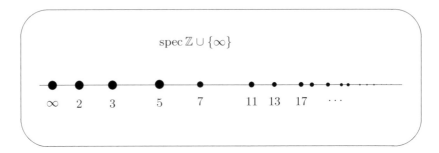

Figure 1.2 spec $\mathbb{Z}$ completed with the archimedean valuation.

In a similar way, the first insight into the valuations of $\mathbb{Q}$ is provided by the archimedean valuation $v_\infty(x) = -\log|x|$. The p-adic valuations are then defined in a manner analogous to (1.10). See Figure 1.2 for an impression of the curve spec $\mathbb{Z}$. Each prime number should correspond to $\log p$ points on this line, and ∞ corresponds to zero, or rather an infinitesimal number of points. The reader may enjoy working on the following exercise, which will deepen his or her understanding of the foregoing proof.

Exercise 1.3.3 Classify the valuations of $\mathbb{Q}$. This is known as Ostrowski's theorem, see [Ost].

Remark 1.3.4 $\mathbb{F}_q(T)$ is isomorphic to the field $\mathbb{F}_q(1/T) = \mathbb{F}_q(S)$. Under this correspondence, the valuation v_∞ corresponds to the valuation v_S (giving the order of vanishing of a function at $S = 0$). Thus we can write

$$v_\infty = v_{1/T},$$

and this valuation corresponds to the point $(0 : 1)$, the point at infinity on the projective line $(S : T)$. See also Section 5.2.

1.4 Global fields

A global field is a finite extension of $\mathbb{Q}$ or a function field with finite field of constants. The completion of the global field K in the valuation v is denoted K_v. In [Art2], the theory of these fields is developed axiomatically as those that satisfy the "product formula."

Lemma 1.4.1 *Let m be a separable irreducible polynomial over K defining an extension $L = K[X]/(m)$ and let v be a valuation of K. Then there is a one-to-one correspondence between irreducible factors of m over K_v and valuations of L extending v.*

Proof Let $m = m_1 m_2 \cdots m_r$ be the factorization of m over K_v and let m_i be a factor. By Lemma 1.2.9, $K_v[X]/(m_i)$ has a unique valuation. Since L embeds in this field, this gives a valuation on L. Thus to every factor of m over K_v there corresponds a valuation of L extending v.

Conversely, let w be a valuation of L extending v. We have a multiplicative map

$$K[X] \longrightarrow L \xrightarrow{\ w\ } \mathbb{R} \cup \{\infty\}, \text{ given by } X \mapsto X + (m) \mapsto w(X + (m)).$$

A polynomial $f \in K[X]$ maps to ∞ if and only if m divides f. By continuity, we obtain a map $K_v[X] \longrightarrow L_w \xrightarrow{\ w\ } \mathbb{R} \cup \{\infty\}$. Since $w(m) = \infty$, it follows that $w(m_i) = \infty$ for some i. If $w(m_j) = \infty$ for another index j, then we observe that m_i and m_j are relatively prime to write $1 \in K$ as a linear combination of m_i and m_j, and obtain $w(1) = \infty$. This contradiction shows that i is unique. Thus to every valuation of L extending v there corresponds an irreducible factor of m over K_v. □

Notation 1.4.2 Let m be a separable irreducible polynomial over K defining an extension $L = K[X]/(m)$ and let v be a valuation of K. The *factor of m corresponding to the valuation w of L extending v* is denoted m_w.

Corollary 1.4.3 *For a separable extension of global fields,*

$$\sum_{w|v} e(w/v) f(w/v) = [L : K]$$

for each valuation v of K.

Proof By Lemma 1.2.12, the degree of m_w is $e(w/v)f(w/v)$. By the proof of the last lemma, the sum of these degrees is the degree of m. □

Example 1.4.4 Consider $K = \mathbf{q}[X]/(m)$, where $m(X) = X^2 - T$ over $\mathbf{q} = \mathbb{F}_3(T)$. The valuation v_T ramifies in K. Indeed, the element T is a uniformizer for v_T, and $T = X^2$ in K. The other ramified valuation is v_∞. For both these valuations, m remains irreducible in $\mathbf{q}_v$.

For an irreducible monic polynomial $P(T) \neq T$, the completion $\mathbf{q}_P$ contains a root of m if T is a square modulo P. By quadratic reciprocity [Ros, Theorem 3.3],

$$\left(\frac{T}{P}\right) = (-1)^{\deg P}\left(\frac{P}{T}\right) = (-1)^{\deg P}P(0).$$

Thus, for example, the valuations v_{T+1} and v_{T^2+T-1} remain prime, and the valuations v_{T-1} and v_{T^2+1} split into two valuations.

Theorem 1.4.5 *All extensions of v to L are obtained from embeddings of L into K_v^a.*

Proof By Corollary 1.2.10, the algebraic closure K_v^a has a valuation extending v, denoted by v^a. Let $\sigma\colon L \to K_v^a$ be an embedding. Then $v^a \circ \sigma$ is a valuation of L extending v.

Conversely, let w be a valuation on L. Then $\sigma\colon L \to K_v[X]/(m_w)$ is an embedding such that $w = v^a \circ \sigma$. □

Theorem 1.4.6 *The trace is the sum of the local traces:*

$$\mathrm{Tr}_{L/K}(x) = \sum_{w|v}\mathrm{Tr}_{w/v}(x)$$

for $x \in L$. Here, $\mathrm{Tr}_{w/v}$ denotes the trace from L_w to K_v.

Proof Take K_v^a as a splitting field for m. Let $x = X + (m)$ be the root of m in L, and let $\sigma_1, \ldots, \sigma_n$ be the embeddings of L in K_v^a. Also, let $m = \prod_{w|v} m_w$ be the factorization of m over K_v. Since by Proposition 1.1.4, $\mathrm{Tr}_{L/K}(y) = \sum_{i=1}^{n}\sigma_i y$ and m splits as $\prod_{i=1}^{n}(X - \sigma_i x)$, we can group together the images $\sigma_i x$ according to the factor m_w that $\sigma_i x$ is a root of. Thus

$$\mathrm{Tr}_{L/K}(y) = \sum_{w|v}\sum_{m_w(\sigma_i x)=0}\sigma_i y.$$

Again by Proposition 1.1.4, the inside sum equals $\mathrm{Tr}_{w/v}(y)$. □

1.4.1 Constants and nonconstant functions

Consider a finite extension of $\mathbf{q} = \mathbb{F}_q(T)$,

$$K = \mathbf{q}[X]/(m).$$

If we do not require m to be monic, then we can clear denominators and assume that the coefficients are polynomials in T without common factor. Then

$$m(T, X) = 0$$

is also an equation for T, and K is a finite extension of $\mathbb{F}_q(X)$. Below, we apply this reasoning to a general element $x \in K$. But how can we be sure that x is not a constant?

We assume that $\mathbb{F}_q$ is algebraically closed in K. This means that if x is an element of K and $x \notin \mathbb{F}_q$, then x is transcendental over $\mathbb{F}_q$.

Lemma 1.4.7 *Let $x \in K \setminus \mathbb{F}_q$. Then K is a finite extension of $\mathbb{F}_q(x)$.*

Proof Since K is finite-dimensional over $\mathbf{q}$ (of dimension $d = \deg_X m$), there exists a relation between the powers $1, x, x^2, \ldots, x^d$ with coefficients in $\mathbf{q}$. Take such a relation of minimal degree in x, and clear denominators and common factors of the coefficients to obtain an irreducible relation

$$f(T, x) = 0.$$

This relationship clearly involves x. It also involves T, because otherwise $f(T, x) = f(0, x) = 0$ would be an equation for x over $\mathbb{F}_q$, contrary to our assumption. It follows that $\mathbb{F}_q(x, T) = \mathbf{q}[x]$ is finite-dimensional over $\mathbb{F}_q(x)$. Since K is finite-dimensional over $\mathbf{q}$, it is finite over $\mathbf{q}[x]$. Hence K is finite over $\mathbb{F}_q(x)$. $\qquad\square$

For the next theorem, see also [Iw, Lemma 7, p. 342] and [Lo].

Theorem 1.4.8 *Let x be a nonvanishing function in K. Then $v(x)$ vanishes for all but finitely many valuations of K. If $v(x)$ vanishes for every valuation of K, then $x \in \mathbb{F}_q$.*

Proof If x is not constant, then x is transcendental over $\mathbb{F}_q$ and K is a finite extension of $\mathbb{F}_q(x)$. This field has the valuations v_x and v_∞ that do not vanish at x. Each extension of these valuations to K is nontrivial at x, but every other valuation vanishes at x. $\qquad\square$

Example 1.4.9 For $x = T(T-1)$ we obtain $\mathbf{q}$ as an extension of $\mathbb{F}_q(x)$ of degree 2. The valuation v_x splits in two valuations v_T and v_{T-1}. The valuation at infinity remains prime.

The function x has only one zero and one pole in $\mathbb{F}_q(x)$. Each extension of v_x to K is a zero of x, and the multiplicity of this zero is the order of ramification. Moreover, the degree of the residue field extension is the number of points in which v_x splits. Thus $\sum_{w|v_x} e(w/v_x) f(w/v_x)$ counts the points above v_x, each with their proper multiplicity. Similarly, each extension of v_∞ is a pole of x. By Corollary 1.4.3, we have the following comparison between the zeros and poles of x:

Corollary 1.4.10   *For an extension K of $\mathbb{F}_q(x)$,*

$$\sum_{w|v_x} e(w/v_x)f(w/v_x) = \sum_{w|v_\infty} e(w/v_\infty)f(w/v_\infty).$$

Both sums equal the degree of K over $\mathbb{F}_q(x)$.

1.4.2 Ramification

The ring of functions that are regular at each valuation of L extending v is

$$\mathfrak{R}_v = \{x \in L \colon w(x) \geq 0 \text{ for every extension } w \text{ of } v \text{ to } L\}. \qquad (1.12)$$

For each w extending v, $\mathfrak{R}_v$ has a maximal ideal $\{x \in \mathfrak{R}_v \colon w(x) > 0\}$.

Lemma 1.4.11 $\mathrm{Tr}_{L/K}(x) \in \mathfrak{o}_v$ *for every $x \in \mathfrak{R}_v$.*

Proof   Extend v to a valuation of K_v^a, denoted by v^a. By Theorem 1.4.5, every extension of v to L is obtained from an embedding $\sigma\colon L \to K_v^a$. Let $x \in \mathfrak{R}_v$. Then $v^a(\sigma x) \geq 0$ for every embedding of L. Since v is nonarchimedean, it follows that $v(\mathrm{Tr}_{L/K}(x)) = v\left(\sum_\sigma \sigma x\right) \geq 0$. $\square$

Definition 1.4.12   The *inverse different at v* of the extension L/K is

$$\mathfrak{d}_{v,L}^{-1} = \left\{x \in L \colon \mathrm{Tr}_{L/K}(xy) \in \mathfrak{o}_v \text{ for every } y \in \mathfrak{R}_v\right\}.$$

Clearly, the inverse different is an $\mathfrak{R}_v$-module containing $\mathfrak{R}_v$.

Lemma 1.4.13   *The different is the product of the local differents:*

$$\mathfrak{d}_{v,L}^{-1} = \left\{x \in L \colon x \in \mathfrak{d}_{w/v}^{-1} \text{ for every extension of } v \text{ to } L\right\}.$$

Proof   Suppose $x \in L$ lies in $\mathfrak{d}_{w/v}^{-1}$ for each $w \mid v$. Let $y \in \mathfrak{R}_v$. Then $\mathrm{Tr}_{w/v}(xy) \in \mathfrak{o}_v$ for each $w \mid v$. By Theorem 1.4.6, it follows that $x \in \mathfrak{d}_{v,L}^{-1}$.

Conversely, if $x \in \mathfrak{d}_{v,L}^{-1}$ then for every $y \in \mathfrak{R}_v$ we have $\mathrm{Tr}_{L/K}(xy) \in \mathfrak{o}_v$. Let w be an extension of v and let $y \in \mathfrak{o}_w$. Approximate y by $y' \in \mathfrak{R}_v \cap \mathfrak{o}_w$ such that $w'(y') \geq \max\{0, -w'(x)\}$ for $w' \neq w$. Then $\mathrm{Tr}_{w'/v}(xy') \in \mathfrak{o}_v$ for each $w' \neq w$. Since $\mathrm{Tr}_{L/K}(xy') \in \mathfrak{o}_v$, we conclude by Theorem 1.4.6 that $\mathrm{Tr}_{w/v}(xy') \in \mathfrak{o}_v$. It follows that $\mathrm{Tr}_{w/v}(xy) \in \mathfrak{o}_v$. Hence $x \in \mathfrak{d}_{w/v}^{-1}$ for every valuation w extending v. $\square$

From Theorem 1.2.17, we deduce the following lemma:

Lemma 1.4.14   *The valuations v of K that ramify in L are exactly those for which $\mathfrak{d}_{v,L} \neq \mathfrak{R}_v$.*

In Chapter 3 we will define the canonical divisor of a curve, which will involve all ramified and infinite valuations. By the next theorem, this is a finite set.

Theorem 1.4.15   *Only finitely many valuations of K ramify in a finite separable extension.*

Proof   Let $L = K[X]/(m)$ be a finite separable extension, defined by a separable monic polynomial m. Let $x = X + (m)$. By Theorem 1.4.8, there are only finitely many valuations such that m does not have coefficients in $\mathfrak{o}_v$ or $v(\det D_x) \neq 0$. By the next lemma, for each extension of every other valuation we have $\mathfrak{o}_v[x] \subseteq \mathfrak{o}_w$ and D_x is invertible modulo π_v. By Theorem 1.2.17, those valuations do not ramify in L. □

Lemma 1.4.16   *Let $m(X) = X^n + m_1 X^{n-1} + \cdots + m_n$ be a monic polynomial defining $L = K[X]/(m)$ and let v be a valuation of K. Let x be a root of m. Then $x \in \mathfrak{o}_w$ for every valuation w of L extending v if and only if the coefficients of m lie in $\mathfrak{o}_v$. Moreover, x is a unit in $\mathfrak{o}_w$ if and only if in addition, m_n is a unit in $\mathfrak{o}_v$.*

Proof   Suppose $m_1, \ldots, m_n \in \mathfrak{o}_v$. Since $x^n = -m_1 x^{n-1} - \cdots - m_n$, we find

$$nw(x) \geq \min\{v(m_i) + (n-i)w(x) \colon i = 1, \ldots, n\}.$$

Let $i \geq 1$ be the index that gives the minimum. Since $v(m_i) \geq 0$, we find that $nw(x) \geq (n-i)w(x)$. Thus $w(x) \geq 0$.

Conversely, since each coefficient of the minimum polynomial of x is a symmetric function of its roots and, by Lemma 1.4.1, if $x \in \mathfrak{o}_w$ for every valuation extending v, then these coefficients lie in $\mathfrak{o}_v$.

To prove that x is a unit if $v(m_n) = 0$, we apply the foregoing to $1/x$, with minimum polynomial $m_n X^n + \cdots + m_1 X + 1$. □

Finally, the following observation is useful for computing the ring of integers of a field (with respect to a choice of valuations at infinity).

Let m be an irreducible monic polynomial with coefficients in $\mathfrak{o}_v$ and let x be a root of m. According to the above lemma, $\mathfrak{o}_v[x] \subseteq \mathfrak{o}_w$ for every extension of v. By the second statement of Lemma 1.2.19, if the matrix D_x is invertible modulo π_v, then $\mathfrak{o}_w = \mathfrak{o}_v[x]$.

We will see that the ring of integers of a function field corresponds to a nonsingular model for the corresponding curve. In Chapter 5, we will compute the desingularization of a curve by an approach using the theorem of Riemann–Roch that does not require us to compute the determinant of the canonical pairing.

2
The local theory

In this and the next chapter, we develop the theory of Iwasawa and Tate of global fields [Iw, Ta] for the case of a function field over a finite field. We only develop the theory with unramified characters, and refer the reader to [vF2] for the theory with ramified characters. We closely follow Tate's exposition, but will omit the abstract introduction of the ring of adeles and its group of units, the ideles, for which [Ta] is an excellent reference.

2.1 Additive character and measure

Let K be a function field with a finite field of constants $\mathbb{F}_q$ and a valuation v. As is well known, its completion K_v is locally compact. It contains the ring $\mathfrak{o}_v$ of function elements regular at v, which has a single maximal ideal $\pi_v \mathfrak{o}_v$ of functions vanishing at v, where π_v is a uniformizer at v. The residue class field $K(v) = \mathfrak{o}_v/\pi_v \mathfrak{o}_v$ is a finite extension of $\mathbb{F}_q$. Its degree over $\mathbb{F}_q$ is denoted by $\deg v$ and we write $q_v = q^{\deg v}$ for its cardinality. We normalize the valuation so that $v(\pi_v) = 1$. The associated norm is

$$|x|_v = q_v^{-v(x)}.$$

We choose a function $T \in K$ such that K is a finite separable extension of $\mathbf{q} = \mathbb{F}_q(T)$. The field K is the field of functions on a curve $\mathcal{C}$, and the choice of T corresponds to choosing a projection of $\mathcal{C}$ onto $\mathbb{P}^1$. In Chapter 5, we study the connection between points on $\mathcal{C}$ and valuations. In particular, we will see that $v(x)$ is the order of vanishing of the function x at any of the points on $\mathcal{C}$ associated with v.

Denote by K_v^+ the additive group of K_v, as a locally compact commutative group, and by x its general element. We wish to determine the character group of K_v^+, and are happy to see that this task is essentially accomplished by the following lemma. We refer to [Ta] for the proof.

Lemma 2.1.1 ([Ta, Lemma 2.2.1])   *If $x \mapsto \chi(x)$ is one nontrivial character of K_v^+, then for each $y \in K_v^+$, $x \mapsto \chi(yx)$ is also a character of K_v^+. The correspondence $y \leftrightarrow \chi(yx)$ is an isomorphism, both topological and algebraic, between K_v^+ and its character group.*

To fix the identification of K_v^+ with its character group promised by this lemma, we must construct a special nontrivial character. We first construct additive characters for $\mathbf{q} = \mathbb{F}_q(T)$. By Section 1.3, the restriction of v to $\mathbf{q}$ is either a multiple of a P-adic valuation for an irreducible polynomial P, or a multiple of v_∞, the valuation at infinity, corresponding to $P(T) = 1/T$. Let $\mathbf{q}_P$ be the completion of $\mathbf{q}$ at P. Thus each element of $\mathbf{q}_P$ is a Laurent series of terms

$$aT^k P^n \quad (a \in \mathbb{F}_q,\ 0 \le k \le \deg P - 1),$$

with only finitely many terms with $n \le 0$.

2.1.1 Characters of $\mathbb{F}_q(T)$

Recall that the characteristic of $\mathbf{q}$ is p. We identify $\mathbb{F}_p$ with $\mathbb{Z}/p\mathbb{Z}$, so that for $n \in \mathbb{F}_p$, the rational number n/p is well defined modulo $\mathbb{Z}$. Define a character χ_P of $\mathbf{q}_P$ as follows. If $P(T) = 1/T$, associated with the valuation at infinity, then

$$\chi_\infty(aT^n) = \begin{cases} \exp\left(-\dfrac{2\pi i}{p}\operatorname{Tr}_{\mathbb{F}_q/\mathbb{F}_p}(a)\right) & \text{if } n = -1, \\ 1 & \text{otherwise}, \end{cases} \qquad (2.1)$$

where $\operatorname{Tr}_{\mathbb{F}_q/\mathbb{F}_p}(a)$ denotes the trace of $a \in \mathbb{F}_q$ over $\mathbb{F}_p = \mathbb{Z}/p\mathbb{Z}$. If, on the other hand, $P(T)$ is an irreducible monic polynomial of degree d with coefficients in $\mathbb{F}_q$, then we put for $0 \le k \le d - 1$ and $a \in \mathbb{F}_q$,

$$\chi_P(aT^k P^n) = \begin{cases} \exp\left(\dfrac{2\pi i}{p}\operatorname{Tr}_{\mathbb{F}_q/\mathbb{F}_p}(a)\right) & \text{if } n = -1 \text{ and } k = d - 1, \\ 1 & \text{otherwise}. \end{cases} \qquad (2.2)$$

Lemma 2.1.2 ([Ta, Lemma 2.2.2]) $x \mapsto \chi_P(x)$ *is a nontrivial, continuous multiplicative map of $\mathbf{q}_P$ into the unit circle group.*

See [Ta] for a proof. It is not true that $\chi_P(x) = 1$ if and only if x is a P-adic integer, unlike the number field case. In fact, χ_P only depends on the coefficient of $1/P$ (or of $1/T$ at the infinite valuation).

Remark 2.1.3 Note the minus sign in the definition of χ_∞. More importantly, note that with $P = 1/T$, $\chi_\infty(aP^n)$ is nontrivial for $n = 1$, whereas for other P, χ_P is nontrivial for $n = -1$. These discrepancies will be resolved by relating the definitions of χ_P and χ_∞ to a residue.

Lemma 2.1.4 *Let P be an irreducible monic polynomial of degree d with coefficients in $\mathbb{F}_q$. Let $\mathbf{q}(P)$ be the residue class field $\mathbb{F}_q[T]/(P)$. Then*

$$\mathrm{Tr}_{\mathbf{q}(P)/\mathbb{F}_q} \left(\frac{T^k}{P'(T)} + (P) \right) = \begin{cases} 0 & \text{if } 0 \le k \le d - 2, \\ 1 & \text{if } k = d - 1. \end{cases}$$

Proof Let $t_1, \ldots, t_d$ be the roots of P in $\mathbf{q}(P)$. Then, by Proposition 1.1.4,

$$\mathrm{Tr}_{\mathbf{q}(P)/\mathbb{F}_q} \left(\frac{T^k}{P'(T)} + (P) \right) = \sum_{i=1}^{d} \frac{t_i^k}{P'(t_i)}.$$

We lift P to a polynomial m over a number field that reduces to P modulo a prime ideal above p with residue class field $\mathbb{F}_q$. The roots of m then reduce to the roots t_i modulo a prime ideal above p in the splitting field of m. Since the roots of P are all distinct, so are those of m. The lemma now follows from the following more general lemma,[1] which asserts that the desired equality already holds without taking the class in $\mathbb{F}_q$. □

Lemma 2.1.5 *Let $m(X)$ be a monic polynomial over $\mathbb{C}$ of degree d without repeated roots. Let $a_1, \ldots, a_d \in \mathbb{C}$ be the roots of m. Then*

$$\sum_{i=1}^{d} \frac{a_i^k}{m'(a_i)} = \begin{cases} 0 & \text{for } 0 \le k \le d - 2, \\ 1 & \text{for } k = d - 1. \end{cases}$$

Proof Consider the contour integral

$$I = \oint_{\Gamma_r} \frac{z^k}{m(z)} \frac{dz}{2\pi i}$$

over the circle of radius r, large enough so that it encircles all the roots of m. By the residue theorem, $I = \sum_{i=1}^{d} a_i^k / m'(a_i)$. On the other hand,

---

[1] Due to Euler. See [Art2, p. 90] or [Sti, Theorem III.5.10, p. 97] for a more elementary proof.

by letting $r \to \infty$, we find that $I = 0$ if $0 \le k \le d - 2$ and $I = 1$ for $k = d - 1$, since $z^k/m(z) = z^{k-d}(1 + O(1/z))$ as $|z| = r \to \infty$. □

Applying Lemma 2.1.4 to definition (2.2), we may write for $n \ge -1$,

$$\chi_P(aT^k P^n) = \exp\left(\frac{2\pi i}{p} \operatorname{Tr}_{\mathbf{q}(P)/\mathbb{F}_p}\left(\frac{P}{P'} aT^k P^n + (P)\right)\right).$$

In this formula, we do not need to assume anymore that P is monic. Note that it also gives the correct value in the case $P = 1/T$ for $n \ge 1$, thus recovering definition (2.1) as well.

Remark 2.1.6   In general, $\oint_{\Gamma_r} z^k m^n(z) \, dz/2\pi i$ vanishes for $n \ne -1$, if Γ_r encircles every root of m. This follows for $n \le -2$ by the same limit argument as in the proof of Lemma 2.1.5, and, for $n \ge 0$, it follows since the integrand is holomorphic. Therefore, we could write the character symbolically as

$$\chi_P(x) = \exp\left(\frac{2\pi i}{p} \operatorname{Tr}_{\mathbb{F}_q/\mathbb{F}_p}\left(\oint_P x \, \frac{dT}{2\pi i}\right)\right),$$

where the notation indicates that all the roots of the polynomial P are to be "encircled." This motivates the following definition:

Definition 2.1.7   For a Laurent series $x = \sum_{i=n}^{\infty} x_i P^i$, with $x_i \in \mathbb{F}_q[T]$ of degree $\deg x_i < \deg P$, the *sum of the residues of x at the points where P vanishes* is defined as

$$\operatorname{res}_P(x) = \operatorname{Tr}_{\mathbf{q}(P)/\mathbb{F}_q}(x_{-1}/P' + (P)).$$

The *residue at infinity* is defined as

$$\operatorname{res}_\infty(x) = -a_1,$$

for a Laurent series $x = \sum_{i=n}^{\infty} a_i T^{-i}$ with coefficients $a_i \in \mathbb{F}_q$,

With this definition, we can simply write, for an irreducible polynomial P or $P = 1/T$,

$$\chi_P(x) = \exp\left(\frac{2\pi i}{p} \operatorname{Tr}_{\mathbb{F}_q/\mathbb{F}_p}(\operatorname{res}_P x)\right).$$

This formula allows easy computation of the values of a character of **q**.

For the following problem, look up the description of the p-adic and real additive characters in [Ta].

Problem 2.1.8   Explain how $-x + \mathbb{Z} \in \mathbb{Q}/\mathbb{Z}$ can be interpreted as the archimedean residue of a rational number x, and how its p-adic fractional part, $x + \mathbb{Z}_p \in \mathbb{Q}_p/\mathbb{Z}_p \subseteq \mathbb{Q}/\mathbb{Z}$, can be interpreted as its residue at p.

It seems to the author that understanding this at a deeper level will give profound information about how to construct the mysterious and elusive field with one element, spec $\mathbb{F}_1$. See also Remark 5.5.1.

2.1.2 Characters of K

Recall that we have chosen a function T such that K is a finite separable extension of $\mathbf{q} = \mathbb{F}_q(T)$. Define for $x \in K_v^+$,

$$\chi_v(x) = \chi_P(\mathrm{Tr}_{v/P}(x)),$$

where v_P is the restriction of v to $\mathbf{q}$ and $\mathrm{Tr}_{v/P}$ denotes the trace from K_v to $\mathbf{q}_P$. Recalling from Theorem 1.1.7 that $\mathrm{Tr}_{v/P}$ is an additive continuous map of K_v onto $\mathbf{q}_P$, we see that χ_v is a nontrivial character of K_v^+. By Lemma 2.1.1, we have proved:

Theorem 2.1.9 *K_v^+ is naturally its own character group if we identify the character $x \mapsto \chi_v(yx)$ with the element $y \in K_v^+$.*

Note that for $P \neq 1/T$, χ_P is trivial on $\mathbf{o}_P$. On the other hand, χ_∞ is trivial on the ideal $T^{-2}\mathbf{o}_\infty$ of $\mathbf{o}_\infty$. By Definition 1.2.16 of the different, we obtain the following lemma:

Lemma 2.1.10 *Let v be a finite valuation. The character $x \mapsto \chi_v(yx)$ associated with y is trivial on $\mathbf{o}_v$ if and only if $y \in \mathfrak{d}_{v/P}^{-1}$, the inverse different of K_v over $\mathbf{q}_P$.*

For an infinite valuation (that is, $v(T) < 0$), the character $x \mapsto \chi_v(yx)$ associated with y is trivial on $x \in \mathbf{o}_v$ if and only if $y \in T^{-2}\mathfrak{d}_{v/\infty}^{-1}$.

Proof We only prove this for a valuation above infinity. Let y be such that $T^2 y \in \mathfrak{d}_{v/\infty}^{-1}$. Then $\mathrm{Tr}_{v/\infty}(T^2 yx) \in \mathbf{o}_\infty$ for $x \in \mathbf{o}_v$. Therefore

$$\mathrm{Tr}_{v/\infty}(yx) \in T^{-2}\mathbf{o}_\infty,$$

so that the character value $\chi_v(yx)$ is 1.

If $y \notin T^{-2}\mathfrak{d}_{v/\infty}^{-1}$, then we can find $x \in \mathbf{o}_v$ such that $\mathrm{Tr}_{v/\infty}(T^2 yx) \notin \mathbf{o}_\infty$. Thus for some $n \leq -1$,

$$\mathrm{Tr}_{v/\infty}(yx) = a_n T^{-n-2} + a_{n+1}T^{-n-3} + \dots, \qquad \text{with } a_n \neq 0.$$

Then $T^{n+1}x \in \mathbf{o}_v$, and $\chi_v(yT^{n+1}x) = \exp\left(-\frac{2\pi i}{p}a_n\right)$ is nontrivial. $\qquad \square$

Recall from (1.9) that $\mathfrak{d}_{v/P}^{-1} = \pi_v^{-d(v/P)}\mathbf{o}_v$. For an infinite valuation, recall from (1.7) the order of ramification $e(v/\infty)$ of v over the infinite valuation of **q**. Define the *canonical exponent*

$$k_v = \begin{cases} d(v/P) & \text{if } v(T) \geq 0, \ v(P) > 0, \\ d(v/\infty) - 2e(v/\infty) & \text{if } v(T) < 0. \end{cases} \qquad (2.3)$$

In particular, $k_P = 0$ for every finite valuation of **q**, and $k_\infty = -2$. In general, $k_v \geq 0$, but above infinity, the canonical exponent may be negative. Moreover, it depends on the choice of the function T in K.

In Section 3.1.2 we will see that $\sum_v k_v v$ is a canonical divisor of the curve $\mathcal{C}$. By Lemma 2.1.10 we obtain:

Theorem 2.1.11 *The character $x \mapsto \chi_v(yx)$ is trivial on $\pi_v^n \mathfrak{o}_v$ if and only if $y \in \pi_v^{-k_v-n} \mathfrak{o}_v$.*

Let μ be a Haar measure for K_v^+. The proof of the following lemma can be found in [Ta, Lemmas 2.2.4 and 2.2.5]. Also see Remark 1.2.3.

Lemma 2.1.12 *For $a \neq 0 \in K_v$ and measurable sets M in K_v^+, define $\mu_1(M) = \mu(aM)$. Then μ_1 is a Haar measure, and consequently there exists a number $\varphi(a) > 0$ such that $\mu_1 = \varphi(a)\mu$. This number equals $|a|_v$, i.e., we have $\mu(aM) = |a|_v \mu(M)$.*

This lemma explains our choice of normalization for the absolute value: the norm $|a|_v$ is the factor by which the additive group K_v^+ is stretched under the transformation $x \mapsto ax$. For the integral, the meaning of the preceding lemma is

$$\int_{K_v^+} f(x)\, \mu(dx) = |a|_v \int_{K_v^+} f(ax)\, \mu(dx).$$

2.1.3 Fourier transform

Let us now select a fixed Haar measure for the additive group K_v^+. Theorem 2.1.9 enables us to do this in an invariant way by selecting the measure which is its own Fourier transform under the interpretation of K_v^+ as its own character group established in that theorem. We state the choice of measure which does this, writing $d_v x$ instead of $\mu(dx)$ and $q_v = q^{\deg v}$. Define

$$d_v x = \text{that measure for which } \mathfrak{o}_v \text{ has measure } q_v^{-k_v/2}. \qquad (2.4)$$

The Fourier transform is defined for integrable functions, and the inversion formula holds for integrable continuous functions for which the Fourier transform is also integrable. See Theorem 2.1.15 below for a sufficient condition for f and $\mathcal{F}_v f$ to be continuous.

Theorem 2.1.13 ([Ta, Theorem 2.2.2]) *If we define the Fourier transform of an integrable continuous function f by*

$$\mathcal{F}_v f(y) = \int_{K_v^+} f(x)\chi_v(yx)\, d_v x, \tag{2.5}$$

then with our choice of measure, the following inversion formula holds:

$$f(x) = \int_{K_v^+} \mathcal{F}_v f(y)\chi_v(-xy)\, d_v y = \mathcal{F}_v \mathcal{F}_v f(-x).$$

We use the notation $\mathbf{p}_v^n$ to denote the indicator function of the set $\pi_v^n \mathbf{o}_v$. By Theorem 2.1.11 and (2.4), the Fourier transform of $\mathbf{p}_v^n$ is

$$\mathcal{F}_v \mathbf{p}_v^n = q_v^{-n-k_v/2} \mathbf{p}_v^{-n-k_v}. \tag{2.6}$$

Example 2.1.14 For the field $\mathbf{q} = \mathbb{F}_q(T)$ of rational functions, the indicator function $\mathbf{p}_P^0$ of the set $\mathbf{o}_P$ is self-dual for every finite valuation v_P. Indeed, for $x = \sum_n x_n P^n$, we have $\chi_P(x) = \chi_P(x_{-1}/P)$. Given $y \notin \mathbf{o}_P$, say

$$y = \sum_{n=-N}^{\infty} y_n P^n \qquad (y_{-N} \neq 0,\ N > 0),$$

we can find an element $x = x_0 P^{N-1} \in \mathbf{o}_P$ such that $\chi_P(yx)$ is not trivial. Therefore, the integral $\int_{\mathbf{o}_P} \chi_P(yx)\, d_P x$ vanishes for $y \notin \mathbf{o}_P$, and by our choice of Haar measure, it equals 1 for $y \in \mathbf{o}_P$. Therefore, $\mathbf{p}_P^0$ is its own Fourier transform.

For the valuation at infinity, we find similarly that the indicator function $\mathbf{p}_\infty^1$ of the maximal ideal $T^{-1}\mathbf{o}_\infty$ is its own Fourier transform.

A function on K_v is *locally constant* at distances in $\pi_v^d \mathbf{o}_v$ if

$$f(x + \varepsilon) = f(x) \text{ for every } \varepsilon \in \pi_v^d \mathbf{o}_v.$$

The Fourier transform interchanges the large and fine structures of a function. This is illustrated by the following lemma:

Theorem 2.1.15 *A function is supported on $\pi_v^n \mathbf{o}_v$ if and only if its Fourier transform is locally constant at distances in $\pi_v^{-k_v - n} \mathbf{o}_v$.*

Proof Suppose f is supported on $\pi_v^n \mathbf{o}_v$. By Theorem 2.1.11, $\chi_v(x\varepsilon)$ is trivial on $\pi_v^n \mathbf{o}_v$ for $\varepsilon \in \pi_v^{-k_v - n} \mathbf{o}_v$. It follows that

$$\mathcal{F}_v f(y + \varepsilon) = \int_{\pi_v^n \mathbf{o}_v} \chi_v(x\varepsilon)\chi_v(xy)f(x)\, d_v x = \mathcal{F}_v f(y).$$

Conversely, if f is locally constant at distances in $\pi_v^{-k_v-n}\mathfrak{o}_v$, then, for every $\varepsilon \in \pi_v^{-k_v-n}\mathfrak{o}_v$,

$$\mathcal{F}_v f(y) = \int_{K_v} \chi_v(xy) f(x-\varepsilon)\, d_v x = \chi_v(\varepsilon y)\mathcal{F}_v f(y).$$

By Theorem 2.1.11, if $y \notin \pi_v^n \mathfrak{o}_v$ then $\chi_v(\varepsilon y) \neq 1$ for some $\varepsilon \in \pi_v^{-k_v-n}\mathfrak{o}_v$. It follows that $\mathcal{F}_v f(y) = 0$, so that $\mathcal{F}_v f$ is supported on $\pi_v^n \mathfrak{o}_v$.   □

We end this section with a well-known property of the Fourier transform. The convolution product of a function f with a function g is defined by

$$C_g f(x) = \int_{K_v} g(x-y) f(y)\, d_v y.$$

The Fourier transform of this function is

$$\mathcal{F}_v C_g f(x) = \iint \chi_v(xz) g(z-y) f(y)\, d_v z\, d_v y.$$

Replacing z by $z + y$, we find

$$\mathcal{F}_v C_g f(x) = \mathcal{F}_v g(x)\mathcal{F}_v f(x).$$

Thus the convolution product is isomorphic to the ordinary product of functions via the Fourier transform,

$$C_g = \mathcal{F}_v^{-1} \circ M_{\mathcal{F}_v g} \circ \mathcal{F}_v,$$

where M_h denotes the multiplication operator $M_h f(x) = h(x)f(x)$.

2.2 Multiplicative character and measure

Our first insight into the structure of the multiplicative group K_v^* of K_v is given by the continuous homomorphism $a \mapsto |a|_v$ of K_v^* into the multiplicative group of powers of $q_v = q^{\deg v}$. The kernel of this homomorphism, the subgroup of all a with $|a|_v = 1$, will obviously play an important role. We denote it by $\mathfrak{o}_v^*$. This group is compact and open.

Concerning the characters of K_v^*, the situation is different from that of K_v^+. Indeed, we are interested in all continuous multiplicative maps of K_v^* into the complex numbers, not only the bounded ones, and shall call such a map a quasi-character, reserving the word "character" for the conventional character of absolute value 1. We call a quasi-character *unramified* if it is trivial on $\mathfrak{o}_v^*$.

Lemma 2.2.1 *The unramified quasi-characters are the maps of the form*

$$|a|_v^s = q_v^{-sv(a)},$$

where s is determined modulo $2\pi i / \log q_v$, $q_v = q^{\deg v}$.

The reader may find in [Ta, Section 2.3] a description of the characters of $\mathfrak{o}_v^*$. We shall only be concerned with unramified quasi-characters.

We will be able to select a Haar measure $d_v^* a$ on K_v^* by relating it to the measure $d_v x$ on K_v^+. If $g(a)$ is integrable on K_v^*, then $g(x)|x|_v^{-1}$ is integrable on $K_v^+ \setminus \{0\}$. So we may define a functional

$$\Phi(g) = \int_{K_v \setminus \{0\}} g(x)|x|_v^{-1} \, d_v x.$$

If $h(a) = g(ba)$ ($b \in K^*$, fixed) is a multiplicative translation of $g(a)$, then

$$\Phi(h) = \int_{K_v \setminus \{0\}} g(bx)|x|_v^{-1} \, d_v x = \Phi(g),$$

as we see by the substitution $x \to b^{-1}x$, $d_v x \to |b|_v^{-1} d_v x$ discussed in Lemma 2.1.12. Therefore our functional Φ, which is obviously nontrivial and positive, is also invariant under translation. It must therefore come from a Haar measure on K_v^*. It will be convenient to have a multiplicative Haar measure that gives the subgroup $\mathfrak{o}_v^*$ measure 1. To this effect we choose as our standard Haar measure on K_v^*:

$$d_v^* a = \frac{q_v^{k_v/2}}{1 - q_v^{-1}} \frac{d_v a}{|a|_v}. \tag{2.7}$$

Since $\mathfrak{o}_v = \mathfrak{o}_v^* \cup \pi_v \mathfrak{o}_v$ is a disjoint union, we obtain the following result:

Lemma 2.2.2 *The measure $d_v^* a$ gives $\mathfrak{o}_v^*$ unit volume.*

We write

$$\kappa_+^* = \frac{q_v^{k_v/2}}{1 - q_v^{-1}} \tag{2.8}$$

for the factor between the multiplicative Haar measure and the additive Haar measure:

$$\int_{K_v^*} f(a) \, d_v^* a = \kappa_+^* \int_{K_v \setminus \{0\}} f(a) \frac{d_v a}{|a|_v}. \tag{2.9}$$

2.3 Local zeta function

The computations are as in [Ta, Section 2.5, p. 319], the $\mathfrak{p}$-adic case. We present here the case when the multiplicative character is unramified.

The local zeta function is the Mellin transform of a function $f \colon K_v \to \mathbb{C}$ on the additive group (see Sections 4.1 and 6.1),

$$\zeta_v(f, s) = \int_{K_v} f(a)|a|_v^s \, d_v^* a. \tag{2.10}$$

This function is periodic with period $2\pi i / \log q_v$, hence also with period $2\pi i / \log q$, independent of v.

The basic example is the function obtained with $f = \mathfrak{p}_v^n$,

$$\zeta_v(\mathfrak{p}_v^n, s) = \int_{\pi_v^n \mathfrak{o}_v} |a|_v^s \, d_v^* a = \sum_{k=n}^{\infty} q_v^{-ks} = \frac{q_v^{-ns}}{1 - q_v^{-s}}. \tag{2.11}$$

Its dual is

$$\zeta_v(\mathcal{F}_v \mathfrak{p}_v^n, s) = q_v^{-n-k_v/2} \frac{q_v^{(n+k_v)s}}{1 - q_v^{-s}}.$$

2.4 Functional equation

The definition of the local zeta function, formula (2.10), is not natural, since we integrate the additive function f against the multiplicative measure. However, letting

$$E_v f(a) = |a|_v^{1/2} f(a) \tag{2.12}$$

be the restriction of f to the multiplicative group, we obtain the local zeta function in a more natural fashion as the Mellin transform of the multiplicative restriction of f,

$$\zeta_v(f, s + 1/2) = \mathcal{M} E_v f(s).$$

Exercise 2.4.1 Let $\mathcal{A}_v$ be the Hilbert space $L^2(K_v, d_v)$ of square-integrable functions on K_v, and let $\mathcal{Z}_v$ be the Hilbert space $L^2(K_v^*, d_v^*)$. Prove that E_v is an isomorphism between $\mathcal{A}_v$ and $\mathcal{Z}_v$ up to the factor $\sqrt{\kappa_+^*}$ from (2.8).

The interplay between the additive Fourier transform and the multiplicative action leads to the functional equation (see also [Ta, p. 314]).

For a function $g\colon K_v \to \mathbb{C}$,

$$\zeta_v(\mathcal{F}_v g, 1 - s) = \int_{K_v^*} \int_{K_v} g(x) \chi_v(bx)\, d_v x\, |b|_v^{1-s}\, d_v^* b.$$

For the product $\zeta_v(f, s) \zeta_v(\mathcal{F}_v g, 1 - s)$ we find, substituting $b \to ab$,

$$\zeta_v(f, s) \zeta_v(\mathcal{F}_v g, 1 - s) = \iint_{K_v^*} \int_{K_v} f(a) g(x) \chi_v(bx) |a|_v^s\, d_v x\, |b|_v^{1-s}\, d_v^* b\, d_v^* a$$

$$= \iint_{K_v^*} \int_{K_v} f(a) g(x) \chi_v(abx) |a|_v\, d_v x\, |b|_v^{1-s}\, d_v^* b\, d_v^* a.$$

By (2.9), we obtain

$$\zeta_v(f, s) \zeta_v(\mathcal{F}_v g, 1 - s) = \kappa_+^* \int_{K_v^*} \iint_{K_v} f(a) g(x) \chi_v(abx)\, d_v x\, d_v a\, |b|_v^{1-s}\, d_v^* b.$$

Since f and g occur symmetrically inside the last expression, we obtain the local functional equation,

$$\zeta_v(f, s) \zeta_v(\mathcal{F}_v g, 1 - s) = \zeta_v(g, s) \zeta_v(\mathcal{F}_v f, 1 - s).$$

It follows that the function

$$\frac{\zeta_v(f, s)}{\zeta_v(\mathcal{F}_v f, 1 - s)} = q_v^{k_v(s-1/2)} \frac{1 - q_v^{s-1}}{1 - q_v^{-s}}$$

is independent of f. See [Ta] for its computation in the archimedean case.

In Chapter 6, this interplay between the additive group and the multiplicative action will be studied in detail.

3

The zeta function

Recall that K is the function field of a curve $\mathcal{C}$ with finite field of constants $\mathbb{F}_q$, where q is a power of p. The ring of *adeles* of K is

$$\mathbb{A} = \mathbb{A}_K = \{x = (x_v) \colon x_v \in K_v \text{ and } x_v \notin \mathfrak{o}_v \text{ for finitely many } v\}.$$

Its multiplicative group is $\mathbb{A}^*$, the group of *ideles*. It is the multiplicative group of vectors (a_v), $a_v \in K_v^*$, such that $a_v \notin \mathfrak{o}_v^*$ for finitely many v.

The ring $\mathbb{A}$ contains the subring of *integral adeles*, and $\mathbb{A}^*$ contains the subgroup of *integral ideles*,

$$\mathfrak{o} = \prod_v \mathfrak{o}_v \subset \mathbb{A}, \quad \text{and} \quad \mathfrak{o}^* = \prod_v \mathfrak{o}_v^* \subset \mathbb{A}^*,$$

respectively. We refer to [Ta, Section 3] for the theory of restricted direct products.

3.1 Additive theory

The ring of adeles of K has the measure $dx = \prod_v d_v x_v$, where the product is taken over all valuations of K. We have an additive fundamental domain for $\mathbb{A}/K$ of unit volume. Note that $\mathbb{A}/K$ can be identified with a subgroup of $\mathbb{A}$, unlike in the number field case. See [Ta, Lemma 4.1.4].

Define a character on $\mathbb{A}$ by

$$\chi(x) = \prod_v \chi_v(x_v)$$

for an adele $x = (x_v)$. By Theorem 1.4.15, the different $\mathfrak{d}_{v/P}$ is nontrivial for only finitely many v. It follows that $\mathbb{A}$ is its own character group via the identification $x \mapsto \chi(yx)$ of an adele y with a character on the adeles.

The Fourier transform of $f \colon \mathbb{A} \to \mathbb{C}$ is defined by

$$\mathcal{F}f(y) = \int_{\mathbb{A}} \chi(yx)f(x)\,dx.$$

By our choice of measure, the inversion formula reads $\mathcal{F}\mathcal{F}f(x) = f(-x)$.

The character χ has the following global property (see Lemma 4.1.5 in [Ta]):

Proposition 3.1.1 $\chi(x) = 1$ *for all* $x \in K$.

This explains our particular choice of local characters. To show this, we decompose a function into partial fractions:

Lemma 3.1.2 *A rational function f can be written as*

$$f(T) = \sum_{P \,:\, v_P(f)<0} \sum_{n=1}^{|v_P(f)|} f_{P,n}(T)P^{-n} + f_\infty(T), \qquad (3.1)$$

where each $f_{P,n}$ is a polynomial in T of degree less than $\deg P$, and f_∞ is a polynomial in T.

Proof Let P be an irreducible polynomial such that $v_P(f) < 0$. Then P is a factor of the denominator of f. Let n be the exponent of P in the denominator. Compute the class of $P^n f$ modulo P. Thus we find a polynomial $f_{P,n}$ of degree less than $\deg P$ such that $f - f_{P,n}P^{-n}$ has the same factors in its denominator as f, to the same power, except that P occurs to a lower power. We continue until $f - \sum f_{P,n}P^{-n}$ has a trivial denominator, and hence is a polynomial $f_\infty(T)$. $\qquad\square$

Example 3.1.3 We compute $\mathbb{A}/\mathbf{q}$ for $\mathbf{q} = \mathbb{F}_q(T)$. Given an adele, for each finite P-adic valuation, we subtract a rational function of the form f/P^k to cancel its denominator. Thus we can subtract an element of $\mathbf{q}$ so that each finite component becomes a regular function in $\mathfrak{o}_P$. Then we can still subtract a polynomial from x to bring the infinite component into $T^{-1}\mathfrak{o}_\infty$. We find $\mathbb{A}/\mathbf{q} = T^{-1}\mathfrak{o}_\infty \times \prod_{P \neq 1/T} \mathfrak{o}_P$.

Since $T^{-1}\mathfrak{o}_\infty$ and each $\mathfrak{o}_P$ has unit volume, $\mathbb{A}/\mathbf{q}$ has unit volume.

Proof of Proposition 3.1.1 We prove this directly for $K = \mathbf{q}$. It then follows in general from the definition of χ_v by Theorem 1.4.6.

Let $x \in \mathbf{q}$ be a rational function. We show that each term in the partial fraction decomposition (3.1) of x contributes trivially to $\chi(x)$.

First, for an irreducible polynomial Q, $\chi_Q(f_\infty) = 1$. Also $\chi_\infty(f_\infty) = 1$ since f_∞ does not have a $1/T$-term. Hence $\chi(f_\infty) = 1$.

Let $f_{P,n}P^{-n}$ be a term in the partial fraction decomposition of x and assume $n \geq 2$. Then $\chi_Q(f_{P,n}P^{-n}) = 1$ for $Q \neq P$ and also for $Q = P$ since $n \geq 2$. Also $\chi_\infty(f_{P,n}P^{-n}) = 1$, since the degree of $f_{P,n}P^{-n}$ is at most $\deg P - 1 - n \deg P \leq -2$.

Finally, a term $f_{P,1}/P$ does not contribute in χ_Q for every irreducible polynomial $Q \neq P$. Assume that P is monic and write

$$f_{P,1}(T) = a_0 + a_1T + \cdots + a_{d-1}T^{d-1},$$

where $d = \deg P$ and $a_i \in \mathbb{F}_q$. The contribution of χ_P is obtained from the trace over $\mathbb{F}_p$ of a_{d-1}. Further, for $i \leq d - 2$, $\chi_\infty(a_iT^i/P) = 1$ because this rational function has degree ≤ -2. Also, since P is monic, $a_{d-1}T^{d-1}/P = a_{d-1}/T + O(T^{-2})$. Hence the contribution of χ_∞ is obtained from the trace over $\mathbb{F}_p$ of $-a_{d-1}$, which cancels the contribution of χ_P. It follows that $\chi(x) = 1$. $\qquad\square$

As in [Ta, Theorem 4.1.4], we deduce the following theorem:

Theorem 3.1.4 *The group $K^\perp$, the characters that are trivial on K, coincides with K.*

Proof By Proposition 3.1.1, K is contained in $K^\perp$. Moreover, $K^\perp$ is clearly a vector space over K. By general Fourier theory, $K^\perp$ is the character group of $\mathbb{A}/K$, which is compact, hence $K^\perp$ is discrete. Further, $K^\perp/K$ is a subgroup of $\mathbb{A}/K$, hence it is compact. Being discrete and compact, $K^\perp/K$ is finite. Since K is infinite, $K^\perp$ must be one-dimensional over K. We conclude that $K^\perp = K$. $\qquad\square$

See Section 3.5 for the S-local adeles, which do not have this property.

3.1.1 Divisors

A *divisor* is a formal sum $D = \sum_v D_v v$, with $D_v \in \mathbb{Z}$. The *degree* of D is $\deg D = \sum_v D_v \deg v$.

The *canonical divisor* of $\mathcal{C}$ is

$$\mathcal{K} = \sum_v k_v v.$$

We write $\deg \mathcal{K} = 2g - 2$, where the *genus* of $\mathcal{C}$ (or of K) is defined as

$$g = 1 + \frac{1}{2}\deg \mathcal{K}.$$

In the next section, when we study the imbedding of K^* into the ideles, divisors will play an important role. Here, they provide us with a convenient notation: $\pi^D \mathfrak{o}$ and $\mathfrak{p}^D$ denote the set $\prod_v \pi_v^{D_v} \mathfrak{o}_v$ and its indicator function, respectively: $\mathfrak{p}^D(a) = 1$ if and only if $(a) \geq D$ (see (3.3) below). By (2.6), we can then write

$$\mathcal{F}\mathfrak{p}^D = q^{1-g-\deg D} \mathfrak{p}^{-K-D}. \tag{3.2}$$

The Fourier transform interchanges the small and large scale of a function. The proof is analogous to the proof of Theorem 2.1.15 and depends on the global counterpart of Theorem 2.1.11:

Theorem 3.1.5   *The character $x \mapsto \chi(yx)$ is trivial on $\pi^D \mathfrak{o}$ if and only if $y \in \pi^{-K-D} \mathfrak{o}$.*

The global counterpart of Theorem 2.1.15 is then:

Lemma 3.1.6   *A function is supported on $\pi^D \mathfrak{o}$ if and only if its Fourier transform is locally constant at distances in $\pi^{-K-D} \mathfrak{o}$.*

The function $D \mapsto \pi^D = \prod_v \pi_v^{D_v}$ associates an idele with each divisor. Conversely, an idele a corresponds to the divisor

$$(a) = \sum_v v(a_v)v \tag{3.3}$$

of degree $\sum_v v(a_v) \deg v$. We see that the group of divisors is isomorphic to $\mathbb{A}^*/\mathfrak{o}^*$. A function $x \in K^*$ gives a *principal divisor*

$$(x) = \sum_v v(x)v.$$

We define the norm of an idele by

$$|a| = \prod_v |a_v|_v,$$

so that $\deg(a) = -\log_q |a|$. By Corollary 1.4.10, a function has as many zeros as poles: $\deg(x) = 0$, or equivalently,

$$|x| = 1 \tag{3.4}$$

for $x \in K^*$ (see also [W8, Theorem IV.4.5] or [Iw, Lemma 7]).

3.1.2 Riemann–Roch

From the Poisson summation formula $\sum_{x \in K} \mathcal{F}f(x) = \sum_{x \in K} f(x)$ we deduce the theorem of Riemann–Roch,

$$\sum_{x \in K} f(xa) = \frac{1}{|a|} \sum_{x \in K} \mathcal{F}f(x/a), \tag{3.5}$$

for every idele a. See also [Ta, Lemma 4.2.4 and Theorem 4.2.1].

Remark 3.1.7 In order that $\sum_K f(xa)$ be defined, it is necessary that f be continuous. This is the case if f is locally constant. Likewise, $\mathcal{F}f$ needs to be continuous. By Lemma 3.1.6, this is the case if f is compactly supported.

To see the geometric significance of the theorem of Riemann–Roch, we take $f = \mathfrak{p}^0$, the indicator function of $\mathfrak{o}$. Let $D = \sum_v D_v v$ be a divisor and let $\pi^D = (\pi_v^{D_v})$ be the corresponding idele. Then

$$\mathfrak{p}^0\left(x\pi^D\right) = \prod_v \mathfrak{p}_v^0\left(\pi_v^{v(x)+D_v}\right) = \begin{cases} 1 & \text{if } (x) + D \geq 0, \\ 0 & \text{otherwise,} \end{cases} \tag{3.6}$$

where a divisor is called *positive* if all its coefficients are positive. The functions $x \in K$ with $(x) + D \geq 0$ form a vector space over $\mathbb{F}_q$,

$$L(D) = \{x \in K : (x) + D \geq 0\}.$$

It is the vector space of functions having pole divisor bounded by D. (It contains the zero function since $v(0) = \infty$ for every valuation.) We denote its dimension over $\mathbb{F}_q$ by $l(D)$.

Exercise 3.1.8 The rest of this argument proves in particular that $l(D)$ is finite.

By (3.6), $\sum_{x \in K} \mathfrak{p}^0\left(x\pi^D\right) = q^{l(D)}$. According to (3.5), we obtain

$$q^{l(D)} = |\pi^{-D}| \sum_{x \in K} \mathcal{F}\mathfrak{p}^0\left(x\pi^{-D}\right).$$

By equation (3.2), $\mathcal{F}\mathfrak{p}^0 = q^{1-g}\mathfrak{p}^{-\mathcal{K}}$. Therefore

$$\mathcal{F}\mathfrak{p}^0\left(x\pi^{-D}\right) = \begin{cases} q^{1-g} & \text{if } (x) - D \geq -\mathcal{K}, \\ 0 & \text{otherwise.} \end{cases}$$

We find $\sum_{x \in K} \mathcal{F}\mathfrak{p}^0(x\pi^{-D}) = q^{1-g}q^{l(\mathcal{K}-D)}$. Since $|\pi^{-D}| = q^{\deg D}$, we obtain the theorem of Riemann–Roch in the classical formulation:

Theorem 3.1.9 *Let $\mathcal{C}$ be a curve of genus g with field of constant functions $\mathbb{F}_q$. Let $\mathcal{K}$ be a canonical divisor on $\mathcal{C}$ and let D be a divisor. Consider the vector space over $\mathbb{F}_q$ of functions f such that $(f) + D$ is positive. Its dimension over $\mathbb{F}_q$ is given by*

$$l(D) = \deg D + 1 - g + l(\mathcal{K} - D).$$

It follows that g is an integer (hence the degree of $\mathcal{K}$ is even). For the trivial divisor we have $L(0) = \mathbb{F}_q$ and hence $l(0) = 1$. Putting $D = 0$ in the Riemann–Roch formula, we find $g = l(\mathcal{K})$, so that $g \geq 0$.

See [Art2, Sti] for the connection with differentials on $\mathcal{C}$.

Example 3.1.10   By the computation of k_v for $\mathbf{q}$, we see that the canonical divisor of $\mathbb{P}^1$ has degree -2, hence the projective line has genus 0. In Lemma 3.6.2 below, we prove that every curve of genus 0 is isomorphic to $\mathbb{P}^1$.

Example 3.1.11   For the extension $K = \mathbf{q}\big[\sqrt{T}\,\big]$ of $\mathbf{q} = \mathbb{F}_3(T)$, we have $d(v/P) = 0$ for each extension of a P-adic valuation for $P \neq T, 1/T$. There is one valuation above v_T and one above $v_{1/T}$, which we denote by $v_{\sqrt{T}}$ and $v_{1/\sqrt{T}}$, respectively. We have $k_{\sqrt{T}} = 1$ and for the infinite valuation, $k_{1/\sqrt{T}} = -3$ (since $e(v_{1/\sqrt{T}}/v_{1/T}) = 2$). Thus the canonical divisor $\mathcal{K} = v_{\sqrt{T}} - 3v_{1/\sqrt{T}}$ has degree -2. We find that the curve $X^2 = T$ has genus 0.

Example 3.1.12   A curve of genus 1 always has a rational point. Indeed, by Theorem 3.3.5 below, we can find a divisor D of degree one. By Riemann–Roch, $l(D) = 1$, hence D is linearly equivalent (see Definition 3.2.1 below) to a positive divisor of degree one, which is a point on $\mathcal{C}$.

Similarly, a curve of genus 0 always has a rational point.

3.2 Multiplicative theory

The multiplicative Haar measure on $\mathbb{A}^*$ is

$$d^*a = \prod_v d_v^* a_v.$$

By Lemma 2.2.2, this measure gives $\mathfrak{o}^*$ unit volume.

3.2.1 Divisor classes

Definition 3.2.1   Two divisors D and D' are *linearly equivalent* if their difference $D - D'$ is the divisor of a function on $\mathcal{C}$.

Thus $D + (\alpha)$ gives all divisors equivalent to D, for $\alpha \in K^*$. By Corollary 1.4.10, if D and D' are linearly equivalent then these divisors have the same degree. We write $\mathrm{Cl}(n)$ for the set of linear equivalence classes of divisors of degree n. The set $\mathrm{Cl}(0)$ is a group, the *divisor class group* of C. Let $\mathcal{P}$ be a divisor of degree n. Then $D \mapsto \mathcal{P} + D$ gives a bijection between the classes of vanishing degree and those of degree n. Thus if $\mathrm{Cl}(n)$ is not empty, then the number of classes in $\mathrm{Cl}(n)$ equals that of $\mathrm{Cl}(0)$.

In Theorem 3.3.5 below, we show that there exist divisors of every degree, so $\mathrm{Cl}(n)$ is never empty. We write

$$h = h_{\mathcal{C}} = |\mathrm{Cl}(0)|$$

for the *class number* of C over $\mathbb{F}_q$.

Example 3.2.2 Let C be a curve of genus 0 and let D be a divisor of degree 0. By the Riemann–Roch formula, $l(D) \geq 1$, hence there exists a function f such that $(f) + D \geq 0$. Since the degree of this divisor vanishes, we have that $D = (1/f)$. This shows that $h = 1$.

In general, let D be a divisor of degree 0 such that $l(D) \geq 1$. Then there exists a function f such that $D + (f) \geq 0$, so that D is linearly equivalent to the trivial divisor. Then $l(D) = 1$, since nonconstant functions always have poles. On the other hand, $l(D) = 0$ for every nontrivial divisor class of vanishing degree.

Theorem 3.2.3 *The number of linear equivalence classes in each degree is finite.*

Proof For $\deg D \geq g$ we have $l(D) \geq 1 + l(K - D) \geq 1$. Hence there exists a function α such that $D + (\alpha) \geq 0$. It follows that the class of a divisor of degree $\geq g$ is represented by a positive divisor. Since there are only finitely many valuations of each degree, the number of elements of $\mathrm{Cl}(n)$ is finite for $n \geq g$. But $|\mathrm{Cl}(n)| = |\mathrm{Cl}(0)|$ if $\mathrm{Cl}(n)$ is not empty, hence h is finite. $\square$

The group of ideles contains the subgroup $\ker |\cdot|$ of ideles of degree zero. This group is a refinement of the group of divisors of degree zero, and $\ker |\cdot|/K^*$ is a refinement of the ideal class group. By (3.4), K^* and $\mathfrak{o}^*$ are subgroups of $\ker |\cdot|$, and their intersection is $\mathbb{F}_q^*$.

To describe the group of idele classes

$$\ker |\cdot|/K^*,$$

we choose ideles $c_1, \ldots, c_h$ of degree zero representing each divisor class. Given an idele of vanishing degree, we can first divide by a c_i to make it the divisor of a function. Then we can divide by that function, determined up to a constant, to make it a unit everywhere. Thus we find a decomposition as a direct product

$$\ker|\cdot| = \bigcup_{c \in \mathrm{Cl}(0)} c \times \mathfrak{o}^* \times K^*/\mathbb{F}_q^*. \tag{3.7}$$

Since $\mathbb{F}_q^*$ contains $q-1$ elements, we find the volume of $\ker|\cdot|/K^*$,

$$\mathrm{vol}(\ker|\cdot|/K^*) = \frac{h}{q-1}. \tag{3.8}$$

Example 3.2.4 Consider the curve $y^2 = x^3 - x$ over $\mathbb{F}_5$. The function x has divisor

$$(x) = -2(\infty) + 2(x = 0, y = 0).$$

Since the curve is not rational, the divisor $-(\infty) + (0,0)$ in $\ker|\cdot|$ represents a torsion point in $\ker|\cdot|/K^*$. We see that $\ker|\cdot|/K^*$ can in general not be identified with a subgroup of $\ker|\cdot|$.

3.2.2 Coarse idele classes

The following map plays an essential role in Chapter 6. It is the combination of a restriction (of a function on the adeles to the ideles) and a trace (the sum over K^*). In that sense, E is a map that "realizes" the noncommutative space $\mathbb{A}/\ker|\cdot|$ in the ordinary space

$$Z = \mathbb{A}^*/\ker|\cdot| \tag{3.9}$$

of *coarse idele classes*. We call it the average restriction.

Definition 3.2.5 We define the *restriction* of f to $\mathbb{A}^*/K^*$ by

$$Rf(a) = \sqrt{|a|} \sum_{\alpha \in K^*} f(\alpha a).$$

If f is $\mathfrak{o}^*$-invariant, the *average restriction* to $\mathbb{A}^*/\ker|\cdot|$ is

$$Ef(a) = \sqrt{|a|} \sum_{\alpha \in \ker|\cdot|/\mathfrak{o}^*} f(\alpha a).$$

Clearly, $R(\mathcal{F}\mathcal{F}f) = Rf$, so that the Fourier transform induces an involution on the image of the restriction map, even though it is not itself an involution. This is because the restrictions of f and $f(-x)$ coincide.

In general, Rf and the restriction of $f(\alpha x)$ coincide, for any $\alpha \in K^*$. Thus the restriction of f only depends on the class of the idele a in $\mathbb{A}^*/K^*$. Further, the average restriction depends only on $\deg a$. As was pointed out above, an element of $\ker |\cdot|$ can be written as $\alpha = c \cdot k \cdot u$ where $c \in \mathrm{Cl}(0)$, $k \in K^*$, and $u \in \mathfrak{o}^*$, and this is almost a direct product, the only ambiguity coming from the fact that $K^* \cap \mathfrak{o}^* = \mathbb{F}_q^*$. Hence

$$Ef(a) = \frac{1}{q-1} \sum_{c \in \mathrm{Cl}(0)} Rf(ca). \tag{3.10}$$

In Section 3.5, we will see that even though the restriction and the average restriction are closely related in the global case, in the semi-local case, the average restriction gives the zeta function as a Mellin transform, whereas the restriction gives Poisson summation and Riemann–Roch (see equation (3.33), Lemma 3.5.9, and the subsequent discussion).

In terms of the restriction, the theorem of Riemann–Roch reads

$$Rf(a) + f(0)|a|^{1/2} = R(\mathcal{F}f)(1/a) + \mathcal{F}f(0)|a|^{-1/2}.$$

By (3.8), it follows that

$$Ef(a) + \frac{h}{q-1}f(0)|a|^{1/2} = E(\mathcal{F}f)(1/a) + \frac{h}{q-1}\mathcal{F}f(0)|a|^{-1/2}. \tag{3.11}$$

Example 3.2.6 Recall the decomposition $\mathbb{A}^* = Z \times \mathrm{Cl}(0) \times \mathfrak{o}^* \times K^*/\mathbb{F}_q^*$ by (3.7) and (3.9). Given a divisor D, we compute $E(\mathfrak{p}^{-D})$:

$$E(\mathfrak{p}^{-D})(a) = \sqrt{|a|} \sum_{\alpha \in \mathrm{Cl}(0) \times K^*/\mathbb{F}_q^*} \mathfrak{p}^{-D}(\alpha a).$$

Now $\mathfrak{p}^{-D}(\alpha a) = 1$ exactly when $(\alpha) + (a) \geq -D$. Such an α can be written as $\alpha = cx$ for some $c \in \mathrm{Cl}(0)$ and $x \in L(c + (a) + D)$. There are $q^{l(c+(a)+D)} - 1$ such x in K^*. Since we need the class of x in $K^*/\mathbb{F}_q^*$ (i.e., the divisor (x)), we obtain

$$E(\mathfrak{p}^{-D})(a) = \sqrt{|a|} \sum_{c \in \mathrm{Cl}(0)} \frac{q^{l(c+(a)+D)} - 1}{q-1}.$$

In Section 3.4, the coefficients of the zeta function will be written in this way.

3.3 The zeta function

As in [Ta], for a function f on the adeles such that f and $\mathcal{F}f$ are of fast decay as $|a| \to \infty$, the zeta function is defined for $\mathrm{Re}\, s > 1$ by

$$\zeta_C(f, s) = \int_{\mathbb{A}^*} f(a)|a|^s \, d^*a. \tag{3.12}$$

In case $f = \prod_v f_v$ is a product of local functions, we find the Euler product of ζ_C by writing the integral over the ideles as a product of local integrals,

$$\zeta_C(f, s) = \prod_v \zeta_v(f_v, s). \tag{3.13}$$

The local factors have been computed in Section 2.3 for $f_v = \mathfrak{p}_v^n$.

As Shai Haran has remarked, definition (3.12) is not natural, since we integrate the additive function f over the multiplicative group $\mathbb{A}^*$ (see also Section 2.4). However, following Haran, we obtain a formula in terms of the restriction by first summing over K^*,[1]

$$\zeta_C(f, s) = \int_{\mathbb{A}^*/K^*} Rf(a)|a|^{s-1/2}\, d^*a. \tag{3.14}$$

If f is $\mathfrak{o}^*$-invariant, we even obtain, by (3.9),

$$\zeta_C(f, s) = \sum_{n \in Z} Ef(n)|n|^{s-1/2}. \tag{3.15}$$

In this sense, $\zeta_C(f, s)$ is the Mellin transform of Ef, in a completely natural way. Since in that case, $f(a)$ only depends on the divisor associated with a, we also obtain

$$\zeta_C(f, s) = \sum_{D \in \mathbb{A}^*/\mathfrak{o}^*} f(D) q^{-s \deg D}. \tag{3.16}$$

Here, we write D for an element of $\mathbb{A}^*/\mathfrak{o}^*$, since this is the group of divisors of $\mathcal{C}$, and we write n for the general element of Z, to remind ourselves that this group is isomorphic to $\mathbb{Z}$ via $n \mapsto \deg n$ (see Section 3.3.1).

Proof Choose a fundamental domain for $\mathbb{A}^*/K^*$ inside $\mathbb{A}^*$ and denote it by $(\mathbb{A}^*/K^*)$. Using $|\alpha| = 1$ for $\alpha \in K^*$, we obtain (3.14) from

$$\zeta_C(f, s) = \sum_{\alpha \in K^*} \int_{(\mathbb{A}^*/K^*)} f(\alpha a)|a|^s\, d^*a.$$

Further, if f is $\mathfrak{o}^*$-invariant, then we use $\mathbb{A}^*/K^* = Z \times \mathrm{Cl}(0) \times \mathfrak{o}^*/\mathbb{F}_q^*$ to write $a = ncu$ for $n \in Z$, $c \in \mathrm{Cl}(0)$ and $u \in \mathfrak{o}^*$. Observing that $\mathfrak{o}^*/\mathbb{F}_q^*$ has volume $\frac{1}{q-1}$, we obtain

$$\zeta_C(f, s) = \frac{1}{q-1} \sum_{n \in Z} \sum_{c \in \mathrm{Cl}(0)} Rf(nc)|n|^{s-1/2},$$

which yields (3.15) by (3.10). $\square$

---

[1] In the proof below, the integral over $\mathbb{A}^*/K^*$ is explained.

Exercise 3.3.1 Define the L-function for a ramified character that is trivial on K^* (i.e., a character of the idele class group $\mathbb{A}^*/K^*$). See [vF2].

To find the analytic continuation, let $\mathbb{A}_n^*$ denote the set of ideles of degree n. Splitting the integral (3.14), the sum over the negative degrees,

$$\sum_{n=1}^{\infty} \int_{\mathbb{A}_{-n}^*/K^*} Rf(a)|a|^{s-1/2} \, d^*a,$$

is finite if f is compactly supported, and in general, it is holomorphic, since we assume f to be of fast decay.

Let t be the minimal positive degree of a divisor and let $\mathbf{t}$ be an idele of norm q^t (for example, for $\mathbf{q}$ we take $\mathbf{t} = (T, 1, 1, \dots)$, with T at the component v_∞). Since the degree is additive, the degree of every divisor is a multiple of t, and we can use $\mathbf{t}^n$ for $n \in \mathbb{Z}$ to represent divisors of each degree.

The sum over the positive degrees,

$$\sum_{n=0}^{\infty} \int_{\mathbb{A}_n^*/K^*} Rf(a)|a|^{s-1/2} \, d^*a,$$

is computed using the theorem of Riemann–Roch. By (3.11), the term of degree tn in this sum equals

$$\int_{\mathbb{A}_{tn}^*/K^*} R\mathcal{F}f(a^{-1})|a|^{s-1/2} \, d^*a + \frac{h}{q-1}\left(\mathcal{F}f(0)q^{tn(1-s)} - f(0)q^{-tns}\right).$$

$$(3.17)$$

The sum of the first terms is holomorphic, since we also assume that $\mathcal{F}f$ is of fast decay, and the sum of the last two terms gives

$$\frac{h}{q-1}\left(\frac{\mathcal{F}f(0)}{1-q^{t(1-s)}} - \frac{f(0)}{1-q^{-ts}}\right).$$

We see that $\zeta_C(f, s)$ has two lines of simple poles, with the following residues: at each point $s = 2k\pi i \,/\, t \log q$ $(k \in \mathbb{Z})$, the residue is

$$-\frac{hf(0)}{t(q-1)\log q},$$

and at $s = 1 + 2k\pi i \,/\, t \log q$, the residue is

$$\frac{h(\mathcal{F}f)(0)}{t(q-1)\log q}.$$

Remark 3.3.2 If f vanishes at 0 then $\zeta_C(f,s)$ has no pole at $s = 0$. This is the case, for example, if $f = \mathfrak{p}^0 - \mathfrak{p}^t$, where t is a divisor of degree one. Likewise, if $\mathcal{F}f(0) = 0$, for example, $f = \mathfrak{p}^0 - q\mathfrak{p}^t$, then $\zeta_C(f,s)$ has no pole at $s = 1$.

The function $f = \mathfrak{p}^0 - \mathfrak{p}^t - q\mathfrak{p}^t + q\mathfrak{p}^{2t}$ satisfies both requirements, so that $\zeta_C(f,s)$ is holomorphic. Note that $\mathfrak{p}^0 - (q+1)\mathfrak{p}^t + q\mathfrak{p}^{2t}$ is like a second difference quotient of $\mathfrak{p}^0$. More precisely, it is a difference with a shift of $\mathfrak{p}^0$ and a dilation (for the L^1-norm; see Section 6.2). See also Exercise 3.4.2.

To finish the computation, we will first show that there exists a divisor of degree one, i.e., $t = 1$, by considering extensions of the field $\mathbb{F}_q$ of constants of K (see [Sti] or [Ei, V.5.3]).

3.3.1 Constant field extensions

We compute the Euler product (3.13), for the special choice $f = \mathfrak{p}^0$. By the computation in Section 2.3, equation (2.11), we obtain

$$\zeta_{C/\mathbb{F}_q}(\mathfrak{p}^0, s) = \prod_v \frac{1}{1 - q_v^{-s}}, \tag{3.18}$$

where the product is taken over all valuations of K.

We denote the constant field extensions of K by $K_n = K[X]/(m)$, where m is an irreducible polynomial over $\mathbb{F}_q$ of degree n. The corresponding zeta function is

$$\zeta_{C/\mathbb{F}_{q^n}}(\mathfrak{p}^0, s) = \prod_w \frac{1}{1 - q_w^{-s}}, \tag{3.19}$$

where w runs over all valuations of K_n.

Let v be a valuation of K. By Lemma 1.4.1, the extensions of v to K_n are found by factoring the polynomial m over K_v. Since for a valuation w of K_n, $K_n(w)$ is the smallest field containing both $K(v)$ and $\mathbb{F}_{q^n}$, we find that $K_n(w)$ has degree $n \deg(v)/d$ over $\mathbb{F}_q$, where $d = \gcd(n, \deg v)$. It follows that the degree of $K_n(w)$ over $K(v)$ is n/d, and that m splits into d different factors of degree n/d. Thus there are d extensions of v to K_n. We find that the local factor corresponding to v in the Euler product for $\zeta_{C/\mathbb{F}_{q^n}}(s)$ is given by

$$\prod_{w|v} \frac{1}{1 - q_w^{-s}} = \left(1 - q_v^{-ns/d}\right)^{-d}.$$

Now $\deg(v)/d$ is relatively prime to n/d. Hence $e^{2\pi i \deg(v)/n}$ is a primitive n/d-th root of unity. We deduce that

$$\left(1 - X^{n/d}\right)^d = \prod_{k=1}^{n} \left(1 - e^{2\pi i k \deg(v)/n} X\right).$$

For $X = q_v^{-s}$, we obtain the following lemma:

Lemma 3.3.3 Let $\zeta_{C/\mathbb{F}_{q^n}}(\mathfrak{p}^0, s)$ be the zeta function of C over the field of constants $\mathbb{F}_{q^n}$, as in (3.19). Then

$$\zeta_{C/\mathbb{F}_{q^n}}(\mathfrak{p}^0, s) = \prod_{k=1}^{n} \zeta_{C/\mathbb{F}_q}\left(\mathfrak{p}^0, s + k \frac{2\pi i}{n \log q}\right).$$

Remark 3.3.4 In Chapter 5, we will see that each valuation v corresponds to an orbit of Frobenius consisting of $\deg v$ points on C with coordinates in $\mathbb{F}_{q^{\deg v}}$. This provides a more geometric way to understand the above argument. Namely, since the Frobenius automorphism of K_n sends a point on C to the point obtained by raising each coordinate to the q^n-th power, the orbit of $\deg v$ points splits into d orbits of length $\deg(v)/d$, where d is as above. Thus there are d extensions of v, and for each extension w we have $\deg w = \deg(v)/d$.

Consider C again over $\mathbb{F}_q$, and let t be the minimal positive degree of a divisor. In particular, the degree of every valuation is a multiple of t. Thus, by (3.18), $\zeta_{C/\mathbb{F}_q}(\mathfrak{p}^0, s)$ has period $2\pi i/t \log q$. By Lemma 3.3.3, we find that $\zeta_{C/\mathbb{F}_{q^t}}(\mathfrak{p}^0, s)$ has a pole of order t at $s = 1$. Since the pole at 1 is always simple, we see that $t = 1$. We conclude that

Theorem 3.3.5 There exists a divisor on C of degree one.

Example 3.3.6 The curve $X^2 + Y^2 + 1 = 0$ has no point over the rational numbers, and every divisor consists of at least two points. More generally, every divisor defined over the rational numbers or the real numbers has an even degree. By the last theorem, we cannot have a similar situation over a finite field. See [V, Chapter 4] for more information on this interesting subject.

We can now obtain the functional equation

$$\zeta_C(f, s) = \zeta_C(\mathcal{F}f, 1 - s), \tag{3.20}$$

by deriving an expression for $\zeta_C(f, s)$ that is symmetric for the substitution $f \leftrightarrow \mathcal{F}f$ and $s \leftrightarrow 1 - s$. By Theorem 3.3.5, t is an idele of degree one and $t = 1$. Substituting $a \to a^{-1}$ in (3.17), we obtain

$$\zeta_C(f, s) = \frac{h}{2(q-1)}\left(f(0)\frac{1+q^s}{1-q^s} + \mathcal{F}f(0)\frac{1+q^{1-s}}{1-q^{1-s}}\right)$$
$$+ \int_{\deg a < 0}\left(Rf(a)|a|^{s-1/2} + R(\mathcal{F}f)(a)|a|^{1/2-s}\right) d^*a$$
$$+ \frac{1}{2}\int_{\deg a = 0}\left(Rf(a) + R(\mathcal{F}f)(a)\right)d^*a,$$

since $Rf(a) + f(0) = R\mathcal{F}f(a) + \mathcal{F}f(0)$ for $\deg a = 0$.

3.3.2 Shifted zeta function

In Chapter 6, it will be convenient to also have the *shifted zeta function*

$$\Lambda_C(f, s) = \zeta_C(f, s + 1/2). \tag{3.21}$$

This function has simple poles at $\pm 1/2 + 2k\pi i/\log q$, for $k \in \mathbb{Z}$, and the Riemann hypothesis states that all its zeros are purely imaginary. If f is $\mathfrak{o}^*$-invariant, $\Lambda_C(f, s)$ is the Mellin transform of Ef, by (3.15),

$$\Lambda_C(f, s) = \sum_{n \in Z} Ef(n)|n|^s.$$

The functional equation for this function is $\Lambda_C(f, s) = \Lambda_C(\mathcal{F}f, -s)$.

3.4 Computation

We compute $\zeta_C(f, s)$ for a particular choice of f. If the canonical divisor is even, i.e., k_v is even for every valuation, then we can take $f = \mathfrak{p}^{-\mathcal{K}/2}$, which is self-dual. Since in general, there may not exist an idele whose square is the canonical divisor, we simply take $f = \mathfrak{p}^0$.[2] We then define the *zeta function* ζ_C *of the curve* C by (3.12), multiplied by the factor $q^{(g-1)s}$,

$$\zeta_C(s) = q^{(g-1)s}\zeta_C(\mathfrak{p}^0, s).$$

By (3.16), this function can be written for $\operatorname{Re} s > 1$ as a sum over positive divisors,

$$\zeta_C(s) = q^{(g-1)s}\sum_{D \geq 0} q^{-s \deg D}.$$

---

[2] By Theorem 3.3.5, we could also take $f = \mathfrak{p}^D$, where D is a divisor of degree $1 - g$, to define $\zeta_C(s) = \zeta_C(f, s)$, without the need of the factor $q^{(g-1)s}$.

By (3.13), we obtain the Euler product over all valuations,

$$\zeta_{\mathcal{C}}(s) = q^{(g-1)s} \prod_v \frac{1}{1 - q_v^{-s}}. \tag{3.22}$$

The factor $q^{(g-1)s}$ has been inserted so that the functional equation (3.20) is self-dual,

$$\zeta_{\mathcal{C}}(s) = \zeta_{\mathcal{C}}(1 - s).$$

We see that $\zeta_{\mathcal{C}}$ is symmetric about $1/2$.

Shifting over $1/2$, we obtain the function (3.21) for this choice of f, with functional equation $\Lambda_{\mathcal{C}}(s) = \Lambda_{\mathcal{C}}(-s)$:

Definition 3.4.1   The *shifted zeta function of* $\mathcal{C}$ is defined by

$$\Lambda_{\mathcal{C}}(s) = \zeta_{\mathcal{C}}(s + 1/2).$$

By the Euler product, for $\operatorname{Re} s > 1/2$ the logarithmic derivative of this function is given by

$$\frac{1}{\log q} \frac{\Lambda'_{\mathcal{C}}}{\Lambda_{\mathcal{C}}}(s) = g - 1 - \sum_v \deg v \sum_{n=1}^{\infty} q_v^{-n(s+1/2)}. \tag{3.23}$$

We now compute $\zeta_{\mathcal{C}}(s)$ more explicitly. For the projective line, of genus zero, each positive divisor without a contribution of the point at infinity corresponds to a monic polynomial. Thus there are q^k such divisors of degree k. Allowing for a contribution of the valuation at infinity, the total number of positive divisors of degree n is $q^n + q^{n-1} + \cdots + 1$ for $n \geq 0$, and for $n < 0$, there are no positive divisors of degree n. Thus the number of positive divisors of degree n is $(q^{n+1} - 1)/(q - 1)$ for $n \geq -1$. We find

$$\zeta_{\mathbb{P}^1}(s) = q^{-s} \sum_{n=-\infty}^{\infty} q^{-ns} \frac{q^{\max\{0,n+1\}} - 1}{q - 1} = \frac{q^{-s}}{(1 - q^{1-s})(1 - q^{-s})}. \tag{3.24}$$

The poles of this function are simple, located at

$$s = k \frac{2\pi i}{\log q}, \quad \text{with residue} \quad -\frac{1}{(q-1)\log q},$$

for $k \in \mathbb{Z}$, and at

$$s = 1 + k \frac{2\pi i}{\log q}, \quad \text{with residue} \quad \frac{1}{(q-1)\log q}.$$

In general, to determine the number of positive divisors of degree n, we count the positive divisors in the linear equivalence class of the divisor D.

Every function $\alpha \neq 0$ in $L(D)$ gives a positive divisor $D + (\alpha)$. And different functions α and β give the same divisor if and only if α/β is a constant in $\mathbb{F}_q^*$. Their number is $(q^{l(D)} - 1)/(q - 1)$. Thus we find for $\mathrm{Re}\, s > 1$,

$$\zeta_{\mathcal{C}}(s) = q^{(g-1)s} \sum_{n=-\infty}^{\infty} q^{-ns} \sum_{D \in \mathrm{Cl}(n)} \frac{q^{l(D)} - 1}{q - 1}. \tag{3.25}$$

For $n < 0$, the degree of D is negative, hence $l(D) = 0$. Thus the sum over n starts at $n = 0$. To make the sum finite, we note that for $n > 2g-2$ (so that $\deg(\mathcal{K} - D) < 0$), we have $l(D) = n + 1 - g$. We find

$$\zeta_{\mathcal{C}}(s) = q^{(g-1)s} \sum_{n=0}^{2g-2} q^{-ns} \sum_{D \in \mathrm{Cl}(n)} \frac{q^{l(D)} - q^{\max\{0, n+1-g\}}}{q - 1} + h\zeta_{\mathbb{P}^1}(s), \tag{3.26}$$

where h is the class number of $\mathcal{C}$. We thus obtain

$$\zeta_{\mathcal{C}}(s) = q^{-gs} L_{\mathcal{C}}(q^s) \zeta_{\mathbb{P}^1}(s), \tag{3.27}$$

where $L_{\mathcal{C}}(X) = X^{2g} + l_1 X^{2g-1} + \cdots + l_{2g-1} X + q^g$ is a polynomial of degree $2g$. From (3.26),

$$L_{\mathcal{C}}(X) = hX^g$$
$$+ (1 - q/X)(X - 1) \sum_{n=0}^{2g-2} X^{2g-1-n} \sum_{D \in \mathrm{Cl}(n)} \frac{q^{l(D)} - q^{\max\{0, n+1-g\}}}{q - 1}. \tag{3.28}$$

We see that $L_{\mathcal{C}}(1) = h$.

The contribution coming from $q^{\max\{0, n+1-g\}}$ to the coefficient of X^n in $L_{\mathcal{C}}(X)$ vanishes for $n \neq g$, and cancels hX^g for $n = g$. Thus the coefficient of X^n is given by

$$l_n = \frac{1}{q - 1} \left(\sum_{D \in \mathrm{Cl}(n)} q^{l(D)} - (q + 1) \sum_{D \in \mathrm{Cl}(n-1)} q^{l(D)} + q \sum_{D \in \mathrm{Cl}(n-2)} q^{l(D)} \right).$$

This formula remains valid for every $n \in \mathbb{Z}$. Indeed, the right-hand side vanishes for $n < 0$ and for $n > 2g$. Compare this formula with Remark 3.3.2.

Exercise 3.4.2 Let D be a divisor of degree one and a an idele of degree n. Show that $l_n = |a|^{-1/2} E(\mathbf{p}^0 - (q+1)\mathbf{p}^D + q\mathbf{p}^{2D})(a)$.

The functional equation for $\zeta_{\mathcal{C}}$ translates to

$$L_{\mathcal{C}}(X) = q^{-g} X^{2g} L_{\mathcal{C}}(q/X).$$

Hence the coefficients satisfy $l_{2g-n} = q^{g-n} l_n$. Therefore, if $l_0, l_1, \ldots, l_g$ have been determined, then the other coefficients can be computed from these.

We write ω_ν for the zeros of $L_{\mathcal{C}}$. Since $L_{\mathcal{C}}$ is monic, we have

$$L_{\mathcal{C}}(X) = \prod_{\nu=1}^{2g} (X - \omega_\nu). \tag{3.29}$$

The analogue of the Riemann hypothesis is that $|\omega_\nu| = q^{1/2}$. We give a proof of this property in Chapter 5.

Example 3.4.3 Let V_n denote the number of valuations of degree n of K. By (3.22), we can compute the logarithmic derivative of $\zeta_{\mathcal{C}}$ to obtain

$$-\frac{1}{\log q} \frac{\zeta_{\mathcal{C}}'}{\zeta_{\mathcal{C}}}(s) = 1 - g + \sum_{n=1}^{\infty} n V_n \frac{q^{-ns}}{1 - q^{-ns}} = 1 - g + \sum_{n=1}^{\infty} q^{-ns} \sum_{d|n} dV_d.$$

On the other hand, by (3.27) and (3.29), we obtain

$$-\frac{1}{\log q} \frac{\zeta_{\mathcal{C}}'}{\zeta_{\mathcal{C}}}(s) = 1 - g - \sum_{\nu=1}^{2g} \frac{\omega_\nu q^{-s}}{1 - \omega_\nu q^{-s}} + \frac{q^{-s}}{1 - q^{-s}} + \frac{q^{1-s}}{1 - q^{1-s}}$$

$$= 1 - g + \sum_{n=1}^{\infty} q^{-ns} \left(q^n + 1 - \omega_1^n - \cdots - \omega_{2g}^n \right).$$

Comparing coefficients, we obtain $\sum_{d|n} dV_d = q^n + 1 - \omega_1^n - \cdots - \omega_{2g}^n$.

For example, for $\mathcal{C} = \mathbb{P}^1$, we have $\sum_{d|n} dV_d = q^n + 1$, from which we obtain

$$V_n = \frac{1}{n} \sum_{d|n} \mu(d) \left(q^{n/d} - 1 \right).$$

See Section 5.4 for the connection with points on $\mathcal{C}$.

3.5 Semi-local theory

For a finite nonempty set S of valuations, the corresponding finite Euler product

$$\zeta_S(s) = \prod_{v \in S} \zeta_v(s)$$

has many poles with $\operatorname{Re} s = 0$, many of high multiplicity if S is large, and no poles with $\operatorname{Re} s = 1$. In particular, ζ_S cannot satisfy a functional equation. Nevertheless, the semi-local theory proceeds quite far. We develop it here to see where it fails.

The ring of S-adeles is

$$\mathbb{A}_S = \prod_{v \in S} K_v.$$

This ring is locally compact with Haar measure $d_S x = \prod_{v \in S} d_v x$. It has an additive character $\chi_S = \prod_{v \in S} \chi_v$, making it its own character group via the identification of $y \in \mathbb{A}_S$ with the character

$$x \mapsto \chi_S(xy).$$

With our choice of measure, the Fourier transform, defined by

$$\mathcal{F}_S f(y) = \int_{\mathbb{A}_S} \chi_S(xy) f(x) \, d_S x,$$

has inverse $\mathcal{F}_S^{-1} f(x) = \mathcal{F}_S f(-x)$.

The field K does not embed discretely in the S-adeles, but the subring of S-integers,

$$\mathcal{O}_S = \{x \in K : v(x) \geq 0 \text{ for } v \notin S\}, \tag{3.30}$$

does lie discretely in $\mathbb{A}_S$. Moreover, the group $\mathbb{A}_S / \mathcal{O}_S$ is compact.

Proposition 3.5.1 *The volume of $\mathbb{A}_S / \mathcal{O}_S$ is $\prod_{v \notin S} q_v^{k_v/2}$.*

Proof Let D denote the divisor $\sum_{v \in S} v$ and let m be such that the degree of mD is at least $2g - 1$. Then $l(mD) = m \deg D + 1 - g$, and for every valuation v, $l(mD + v) = m \deg D + \deg v + 1 - g$. This means that for every $v \in S$, there are $\deg v$ independent functions that have poles only at valuations in S and have a pole in v of order $m + 1$. Since the residue class field $K(v)$ also has degree $\deg v$ over $\mathbb{F}_q$, these $\deg v$ functions can be used to cancel the term of degree $m + 1$ of the pole at v of an adele in $\mathbb{A}_S$.

This shows that every element of $\mathbb{A}_S / \mathcal{O}_S$ is represented by an element of $\pi^{-mD} \mathfrak{o}_S$, where $\mathfrak{o}_S = \prod_{v \in S} \mathfrak{o}_v$. It can then still be modified by an element of $\mathcal{O}_S$ that lies in $L(mD)$.

The volume of $\pi^{-mD} \mathfrak{o}_S$ is

$$\prod_{v \in S} q_v^m \prod_{v \in S} q_v^{-k_v/2} = q^{m \deg D} \prod_{v \in S} q_v^{-k_v/2},$$

and the number of elements in $L(mD)$ is $q^{l(mD)} = q^{m \deg D}q^{1-g}$. The volume of $\mathbb{A}_S/\mathcal{O}_S$ is found as the quotient, using that $g-1 = \sum_v (\deg v)k_v/2$.

$\square$

Exercise 3.5.2 If $S = \{v\}$ contains a single valuation, then $\mathbb{A}_S = K_v$, $\mathfrak{o}_S = \mathfrak{o}_v$, and $\mathcal{O}_S$ is the ring of functions that only have a pole at v. Verify Proposition 3.5.1 in this case.

Let

$$\mathcal{O}_S^\perp = \{y \in \mathbb{A}_S : \chi_S(xy) = 1 \text{ for all } x \in \mathcal{O}_S\}$$

be the subgroup of characters that are trivial on $\mathcal{O}_S$.

Exercise 3.5.3 For $K = \mathbf{q}$, the rational function field, describe $\mathcal{O}_S$ in case S contains the valuation at infinity, and in case S does not contain this valuation. Show that in the first case, $\mathcal{O}_S^\perp = \mathcal{O}_S$, and in the second case, $\mathcal{O}_S^\perp = T^{-2}\mathcal{O}_S$. See also Corollary 3.5.7.

An element of $\mathcal{O}_S$ has its poles only in S, but it may have zeros outside of S, that are then "invisible." Therefore it may have more poles than zeros. On the other hand, the divisor of an element of the group of *S-units*

$$\mathcal{O}_S^* = \{x \in K : v(x) = 0 \text{ for } v \notin S\}, \tag{3.31}$$

has vanishing degree. This group embeds diagonally in the group of *S*-ideles,

$$\mathbb{A}_S^* = \prod_{v \in S} K_v^*.$$

The kernel of the *S*-norm (which is defined by $|\cdot|_S = \prod_{v \in S} |\cdot|_v$) contains the compact subgroup $\mathfrak{o}_S^* = \prod_{v \in S} \mathfrak{o}_v^*$, which is the group of units of $\mathfrak{o}_S$. The quotient $\ker |\cdot|_S/\mathcal{O}_S^*$ is compact.

Exercise 3.5.4 Prove that $\ker |\cdot|_S/\mathcal{O}_S^*$ is compact.

Lemma 3.5.5 $\mathcal{O}_S^\perp$ *is an* $\mathcal{O}_S$-*module, and it contains a copy of* $\mathcal{O}_S$.

Proof By Theorem 2.1.11, for $a \in \pi_v^{-k_v}\mathfrak{o}_v$, the character $x \mapsto \chi_v(ax)$ is trivial on $\mathfrak{o}_v$. Choose a nonzero element $a \in \mathcal{O}_S$ such that

$$a \in \pi_v^{-k_v}\mathfrak{o}_v \quad \text{for all } v \notin S. \tag{3.32}$$

Since an element $x \in \mathcal{O}_S$ lies in $\mathfrak{o}_v$ for all $v \notin S$, it follows that $\chi_v(ax) = 1$ on $\mathcal{O}_S$ for all $v \notin S$. Since the global character $\chi(ax) = 1$ on K, and hence on $\mathcal{O}_S$, we find that $\chi_S(ax) = 1$ on $\mathcal{O}_S$. We have found a nontrivial element $a \in \mathcal{O}_S^\perp$ such that the corresponding character is trivial on $\mathcal{O}_S$. Since $\mathcal{O}_S^\perp$ is clearly a $\mathcal{O}_S$-module, we find the subgroup $a\mathcal{O}_S$ inside this group. $\qquad \square$

Lemma 3.5.6 *The group $\mathcal{O}_S^\perp$ is contained in K. For $a \in \mathcal{O}_S$ as in (3.32), $\mathcal{O}_S^\perp / a\mathcal{O}_S$ is finite.*

Proof Since $\mathcal{O}_S^\perp$ is the character group of the compact group $\mathbb{A}_S / \mathcal{O}_S$, it is discrete. Moreover, multiplication by a is a continuous automorphism of the S-adeles, hence $\mathbb{A}_S / a\mathcal{O}_S$ is compact. Since $\mathcal{O}_S^\perp / a\mathcal{O}_S$ is a discrete subgroup of this compact group, it is finite. However, we can not conclude that $\mathcal{O}_S^\perp = a\mathcal{O}_S$.[3] Instead, given $b \in \mathcal{O}_S^\perp \setminus a\mathcal{O}_S$, we can consider the elements xb for $x \in \mathcal{O}_S$. Since $\mathcal{O}_S$ is infinite, we can find x and x' such that $(x - x')b$ is trivial in $\mathcal{O}_S^\perp / a\mathcal{O}_S$. Thus we find a nonzero $y \in \mathcal{O}_S$ such that $yb \in a\mathcal{O}_S$ (this y may have zeros outside S, so that it is not invertible in $\mathcal{O}_S$). Doing this for each b in a set of representatives, we find that $a\mathcal{O}_S \subseteq \mathcal{O}_S^\perp \subseteq (a/y)\mathcal{O}_S$ for some $y \in \mathcal{O}_S$. $\qquad \square$

From the foregoing proof, we obtain more precise information:

Corollary 3.5.7 $\mathcal{O}_S^\perp = \{y \in K : v(y) \geq -k_v \text{ for all } v \notin S\}.$

Exercise 3.5.8 Prove this corollary.

We define the S-local zeta function by

$$\zeta_S(s) = \int_{\mathbb{A}_S^*} \mathfrak{p}_S^0(x) |x|_S^s \, d_S^* x.$$

We immediately find the Euler product $\zeta_S(s) = \prod_{v \in S} \left(1 - q_v^{-s} \right)^{-1}$.

Recalling that $|a|_S = 1$ for $a \in \mathcal{O}_S^*$, we also obtain an expression in terms of the group of S-idele classes,

$$\zeta_S(s) = \sum_{a \in \mathcal{O}_S^*} \int_{\mathbb{A}_S^* / \mathcal{O}_S^*} \mathfrak{p}_S^0(ax) |x|_S^s \, d_S^* x.$$

As before, we find

$$\zeta_S(1/2 + s) = \mathcal{M} E_S \big(\mathfrak{p}_S^0 \big)(s), \qquad (3.33)$$

where the average S-restriction is $E_S f(n) = \sqrt{|n|_S} \sum_{\alpha \in Z_S} f(\alpha n)$. Note that $Z_S = \ker |\cdot|_S / \mathfrak{o}_S^*$ contains the group $\mathcal{O}_S^* / \mathbb{F}_q^*$.

We recall Poisson summation from abstract Fourier analysis:

---

[3] Compare with Theorem 3.1.4, which discusses the case when S contains all valuations.

Lemma 3.5.9 (Poisson summation)   *For a continuous function with continuous Fourier transform,*

$$\sum_{x \in \mathcal{O}_S} f(x) = \prod_{v \notin S} q_v^{-k_v/2} \sum_{y \in \mathcal{O}_S^\perp} \mathcal{F}_S f(y). \tag{3.34}$$

Proof This follows from the abstract theory. We only need to check it in one example, say $f = \mathfrak{p}_S^0$. The left-hand side equals q, since $f(x)$ is nonzero only for $x \in \mathbb{F}_q$. Since $\mathcal{F}_S \mathfrak{p}_S^0 = \prod_{v \in S} q_v^{-k_v/2} \mathfrak{p}_v^{-k_v}$, the right-hand side is a sum over $y \in \mathcal{O}_S^\perp \subset K$ such that $v(y) \geq -k_v$ for $v \in S$. By Corollary 3.5.7, $v(y) \geq -k_v$ also for $v \notin S$. There are $q^{l(K)} = q^g$ such elements y. □

We see that the Poisson summation formula has a sum over $\mathcal{O}_S$, and if f vanishes at 0, then we still have a sum over the multiplicative semi-group $\mathcal{O}_S^\times = \mathcal{O}_S \setminus \{0\}$. Thus the S-restriction map,

$$R_S f(x) = \sqrt{|x|_S} \sum_{a \in \mathcal{O}_S^\times} f(ax),$$

must be defined using $\mathcal{O}_S^\times$. On the other hand, to have a connection with the S-local zeta function via (3.33), we would want to define the average restriction using the units $\mathcal{O}_S^*$ of $\mathcal{O}_S$. This is a much smaller set than $\mathcal{O}_S^\times$, unless S contains all valuations, in which case both sets coincide with K^*. Because of this discrepancy, we cannot continue our computation of the S-local zeta function, and, in particular, we do not find a functional equation.

Exercise 3.5.10   Describe the explicit formula of the next chapter in the S-local situation.

3.6 Two-variable zeta function

Surprisingly, the zeros of $L_C(X)$ lie in an irreducible algebraic family. In [Pel], Pellikaan defines the following two-variable zeta function (see also [LagR]):

$$\zeta_C(s, t) = \frac{q^{(g-1)s}}{q^t - 1} \sum_{n=-\infty}^{\infty} q^{-ns} \sum_{D \in \mathrm{Cl}(n)} q^{tl(D)}.$$

This series is convergent for $\mathrm{Re}\, t < \mathrm{Re}\, s < 0$.

For $\mathcal{C} = \mathbb{P}^1$, we can compute this function by using that $l(D) = 0$ for $\deg D < 0$, $l(D) = \deg D + 1$ for $\deg D \geq 0$, and $h = 1$. We find

$$\zeta_{\mathbb{P}^1}(s,t) = \frac{1}{(1 - q^{t-s})(q^s - 1)}.$$

Thus $\zeta_{\mathbb{P}^1}(s,t)$ has a meromorphic continuation to the entire (s,t)-plane. Moreover, $\zeta_{\mathbb{P}^1}(t - s, t) = \zeta_{\mathbb{P}^1}(s,t)$ and $\zeta_{\mathbb{P}^1}(s,1) = \zeta_{\mathbb{P}^1}(s)$ as in (3.24). The poles of $\zeta_{\mathbb{P}^1}$ are simple, located in (s,t)-space at the planes $s = 0 + k\frac{2\pi i}{\log q}$ and $s = t + k\frac{2\pi i}{\log q}$, for $k \in \mathbb{Z}$.

In general, we have

$$\zeta_{\mathcal{C}}(s,t) = h\zeta_{\mathbb{P}^1}(s,t) + \sum_{n=0}^{2g-2} q^{-(n+1-g)s} \sum_{D \in \mathrm{Cl}(n)} \frac{q^{t l(D)} - q^{t \max\{0, n+1-g\}}}{q^t - 1},$$

where h is the class number of $\mathcal{C}$. Thus $\zeta_{\mathcal{C}}(s,t)$ also has a meromorphic continuation, with the same poles as $\zeta_{\mathbb{P}^1}(s,t)$. By (3.26), $\zeta_{\mathcal{C}}(s,1) = \zeta_{\mathcal{C}}(s)$. Also, by the theorem of Riemann–Roch, $\zeta_{\mathcal{C}}(t - s, t) = \zeta_{\mathcal{C}}(s,t)$.

We can write

$$\zeta_{\mathcal{C}}(s,t) = q^{-gs}\zeta_{\mathbb{P}^1}(s,t)L_{\mathcal{C}}(q^s, q^t),$$

where

$$L_{\mathcal{C}}(X,Y) = hX^g$$
$$+ (X - Y)(X - 1)\sum_{n=0}^{2g-2} X^{2g-2-n} \sum_{D \in \mathrm{Cl}(n)} \frac{Y^{l(D)} - Y^{\max\{0, n+1-g\}}}{Y - 1}.$$

$$(3.35)$$

In particular,

$$L_{\mathcal{C}}(1, Y) = h. \tag{3.36}$$

Exercise 3.6.1 Compute and interpret $L_{\mathcal{C}}(X, 1)$ and $\zeta_{\mathcal{C}}(s, 0)$.

We see that $L_{\mathcal{C}}(X, Y)$ is obtained from (3.28) by replacing q by Y.[4] Thus

$$L_{\mathcal{C}}(X, Y) = X^{2g} + l_1(Y)X^{2g-1} + \cdots + l_{2g-1}(Y)X + Y^g,$$

---

[4] However, it is not true that $L_{\mathcal{C}}(X)$ determines $L_{\mathcal{C}}(X, Y)$, see [Pel, Example 3.7]. In particular, $L_{\mathcal{C}}(X, Y)$ cannot be computed by counting the number of points on $\mathcal{C}(\mathbb{F}_{q^n})$ for finitely (or infinitely) many values of n. See also Exercise 5.4.1.

where

$$l_n(Y) = \frac{1}{Y-1}\left(\sum_{\text{Cl}(n)} Y^{l(D)} - (Y+1)\sum_{\text{Cl}(n-1)} Y^{l(D)} + Y\sum_{\text{Cl}(n-2)} Y^{l(D)}\right).$$

$$(3.37)$$

Clearly, $l_n(Y)$ is a polynomial in Y. By the functional equation,

$$l_{2g-n}(Y) = Y^{g-n}l_n(Y).$$

Moreover, $l_0(Y) = 1$ and $l_{2g}(Y) = Y^g$.

3.6.1 The polynomial $L_C(X, Y)$

It is remarkable that the polynomial $L_C(X, Y)$ is irreducible. Thus the zeros of $\zeta_C(s)$ lie on an irreducible algebraic curve in $\mathbb{C}^2$ as the fiber above $Y = q$. To prove this, we need a lemma characterizing curves of genus 0.

Lemma 3.6.2 *Let D be a divisor of C of degree $\deg D \geq 1$ and such that $l(D) \geq \deg D + 1$. Then C is isomorphic to $\mathbb{P}^1$.*

Proof Let D be a divisor of degree one such that $l(D) \geq 2$. We can find two independent functions α and β such that $D + (\alpha)$ and $D + (\beta)$ are positive. Since these divisors have degree one, each consists of a single point, say v and w, respectively. Then $(\alpha/\beta) = v - w$. Since α and β are independent, the function α/β is nonconstant from C to $\mathbb{P}^1$, and in particular, $v \neq w$, by Theorem 1.4.8. If $\alpha/\beta(a) = \infty$ then $a = w$. Moreover, if $\alpha/\beta(a) = \alpha/\beta(b) \neq \infty$, then the divisor of $\alpha/\beta - \alpha/\beta(a)$ equals $(a) - w = (b) - w$, hence $a = b$. Thus α/β is an isomorphism of C with $\mathbb{P}^1$.

Let now D be a divisor on C of degree d with $l(D) \geq d+1 \geq 2$. If $d = 1$, then $C = \mathbb{P}^1$ by the first part of the proof. If $d \geq 2$, let then $\alpha_0, \alpha_1, \ldots, \alpha_d$ be $d + 1$ independent functions in $L(D)$ and let v be a valuation where some functions α_i have a pole. Without loss of generality, assume that among the α_i's, the function α_d has a pole of maximum order at v. Then we can find constants λ_i (in the algebraic closure $\mathbb{F}_q^a$) such that $\alpha_i - \lambda_i\alpha_d$ has a pole of lower order at v, for $0 \leq i \leq d - 1$. There are d such functions, they are independent, and they lie in $L(D - v)$. Thus the divisor $D - v$ has degree $d - 1$ and $l(D - v) \geq d$. Continuing this way, we find a divisor D' of degree one with $l(D') \geq 2$. We conclude that $C = \mathbb{P}^1$ by the first part of the proof. $\qquad\square$

Corollary 3.6.3 *If $g \geq 1$ and $\deg D \leq 2g - 3$, then $l(D) \leq g - 1$.*

Proof For $\deg D \geq 1$, $l(D) \leq \deg D$ by the foregoing lemma. It follows that $l(\mathcal{K} - D) = g - 1 - \deg D + l(D) \leq g - 1$. Moreover, $\mathcal{K} - D$ is a divisor of degree at most $2g - 3$. □

Lemma 3.6.4 *Let C have genus $g \geq 1$. Then the top term of $L_C(X, Y)$ as a polynomial in Y is $(1 - X)Y^g$.*

Proof First of all, $l_{2g}(Y) = Y^g$, which gives the term Y^g in $L_C(X, Y)$.

For $n = 2g - 1$, we have $l(D) = g$ for every divisor D of degree n by the theorem of Riemann–Roch, since $\mathcal{K} - D$ has negative degree. Therefore, $\sum_{\text{Cl}(n)} Y^{l(D)} = hY^g$. Also, for a divisor D of degree $n - 1$, we have $l(D) = g - 1 + l(\mathcal{K} - D)$, so $l(D) = g - 1$ for the $h - 1$ divisor classes unequal to $\mathcal{K}$ and $l(D) = g$ for $D = \mathcal{K}$. Hence

$$\sum_{\text{Cl}(n-1)} Y^{l(D)} = (h - 1)Y^{g-1} + Y^g.$$

By formula (3.37), we obtain

$$l_{2g-1}(Y) = Y \sum_{\text{Cl}(2g-3)} \frac{Y^{l(D)} - Y^{g-1}}{Y - 1} - Y^g + (h - 1)Y^{g-1}.$$

By the corollary, the sum is $O(Y^{g-1})$. We find

$$l_{2g-1}(Y) = -Y^g + O(Y^{g-1}).$$

Finally, by (3.37) for $n \leq 2g - 2$, we have that $l_n(Y) = O(Y^{g-1})$. We conclude that $L_C(X, Y) = (1 - X)Y^g + O(Y^{g-1})$. □

We can now prove Naumann's theorem [Nau]. Note that being irreducible in $\mathbb{C}(Y)[X]$ is stronger than irreducibility in $\mathbb{C}[Y, X]$ (and hence equivalent, by application of Gauss's lemma as in the first part of the following proof).

Exercise 3.6.5 Formulate Gauss's lemma. Then prove it.

Exercise 3.6.6 Show that if $L(X, Y)$ is a polynomial in two variables and $L = fg$ is a factorization of L in two polynomials in Y with rational coefficients in X, then L also factorizes in polynomials in X and Y. That is, there is a factorization where the coefficients of the powers of Y are polynomials in X.

Theorem 3.6.7 *The polynomial $L_C(X, Y)$ is irreducible in $\mathbb{C}(Y)[X]$.*

Proof If we have a factorization

$$L_{\mathcal{C}}(X,Y) = f(X,Y)g(X,Y)$$

in polynomials f and g in X with rational coefficients in Y, then we can assume by Gauss's lemma that f and g are polynomials in both X and Y. Since the top coefficient of $L_{\mathcal{C}}(X,Y)$ as a polynomial in Y is irreducible by Lemma 3.6.4, we can assume, without loss of generality, that the top coefficient of f as a polynomial in Y is constant. Thus we have that $\deg_Y f(X,Y) = \deg_Y f(1,Y)$. Since $f(1,Y)g(1,Y) = h_{\mathcal{C}}$ by (3.36), we find that $\deg_Y f(X,Y) = 0$. Thus $f(X,Y) = f(X)Y^0$. Since the top coefficient of f as a polynomial in Y is constant, we find that $f(X)$ is constant. We conclude that $L_{\mathcal{C}}(X,Y)$ is irreducible. □

It is natural to ask about the geometry of the curve $L_{\mathcal{C}}(X,Y) = 0$. For example, what is its genus? The combined degree of each term in the sum in (3.35) is at most

$$2g - \deg D + l(D) - 1 = g + l(K - D).$$

Now $g + l(K - D) = 2g$ for $D = 0$, and $g + l(K - D) \le 2g - 1$ for $\deg D > 0$ or $\deg D = 0$ but $D \not\equiv 0$. Hence the degree of the curve $L_{\mathcal{C}}(X,Y) = 0$ is $2g$. If this curve were nonsingular, then its genus would be $(2g - 1)(g - 1)$, where g is the genus of $\mathcal{C}$.

Problem 3.6.8 Let $\mathcal{C}$ be an algebraic curve and let $L_{\mathcal{C}}(X,Y)$ be the numerator of its two-variable zeta function. What is the genus of the curve defined by $L_{\mathcal{C}}(X,Y) = 0$? Is this curve nonsingular?

Example 3.6.9 The two-variable zeta function of a hyperelliptic curve is determined by its one-variable zeta function [Pel, Proposition 4.3]. For example, the numerator of the two-variable zeta function of a curve of genus 1 is determined by the class number:

$$L_{\mathcal{C}}(X,Y) = X^2 + (h - 1 - Y)X + Y.$$

Since this is linear in Y, the curve $L_{\mathcal{C}}(X,Y) = 0$ is rational and nonsingular.

An example is the elliptic curve

$$X^2 = T^3 - T$$

over $\mathbb{F}_3$. It has four points over $\mathbb{F}_3$, hence its class number is $h = 4$. We find the numerator of its zeta function,

$$L_{\mathcal{C}}(3^s, 3^t) = 3^{2s} + (3 - 3^t)3^s + 3^t.$$

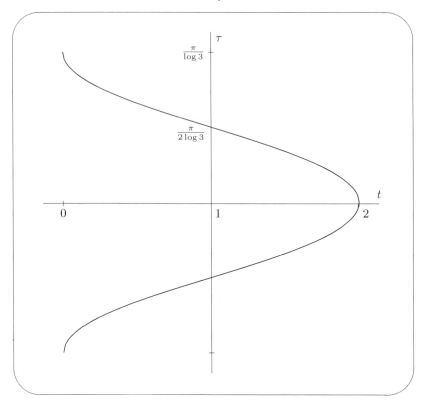

Figure 3.1 An irreducible family of zeros.

For $0 \le t \le 2$, that is, for $1 \le Y \le 9$, the equation $L_{\mathcal{C}}(X, Y) = 0$ has two complex conjugate solutions in X, both of absolute value $\sqrt{Y} = 3^{t/2}$,

$$X = \frac{Y - 3 \pm i\sqrt{(Y-1)(9-Y)}}{2}.$$

Writing $Y = 3^t$ and $X = 3^{t/2+i\tau}$ for a solution of $L_{\mathcal{C}}(X, Y) = 0$, we have plotted the argument τ of s as a function of t in Figure 3.1. In particular, for $t = 1$, we see the two solutions

$$s = \frac{1}{2} \pm \frac{\pi i}{2 \log 3}.$$

For $t < 0$ and $t > 2$, and generally for complex values of t, the real parts of the solutions in s are not equal to $t/2$.

4
Weil positivity

The explicit formula of prime number theory relates the zeros of a zeta function (the spectral side) with the counting function of divisors (the geometric side). It was first discovered by Riemann in 1859 (and probably already earlier), and was then established rigorously by de la Vallée-Poussin in 1896 [dV1, dV2].[1]

The explicit formula was first viewed as a solution to the problem of counting prime numbers. At the time, the spectral side was considered relatively well understood, since it was easily established that the Riemann zeta function does not vanish on the line $\operatorname{Re} s = 1$.

Nowadays, the opposite opinion may be more widely held: the spectral side is not very well understood, and it may be possible to use the explicit formula and Weil positivity to gain insight into the zeros of a zeta function by studying the geometry (of $\mathcal{C}$ or of $\operatorname{spec} \mathbb{Z}$).

4.1 Functions on the coarse idele classes

In Chapter 6, we will introduce the Hilbert space of sequences on the coarse idele class group $Z = \mathbb{A}^*/\ker|\cdot|$. In this section, we consider periodic functions that are analytic in the strip $|\operatorname{Re} s| \leq 1/2$ with period $2\pi i/\log q$. The Fourier series of such a function is

$$f(s) = \sum_{n=-\infty}^{\infty} a(\mathfrak{t}^n) q^{ns},$$

---

[1] Hadamard also published a proof of the prime number theorem in that year [Had], but his approach did not use the explicit formula.

where t is an idele of norm q (see Theorem 3.3.5). We call this the *Mellin transform* of a,

$$\mathcal{M}a(s) = \sum_{n=-\infty}^{\infty} a(\mathsf{t}^n)q^{ns}. \tag{4.1}$$

Remark 4.1.1 The Mellin transform of a function on $(0, \infty)$ is usually defined as

$$\mathcal{M}a(s) = \int_0^\infty a(t)t^s \frac{dt}{t}.$$

We have here the discrete version of this formula.

We can recover $a(\mathsf{t}^n)$ by the inversion formula,

$$a(\mathsf{t}^n) = \frac{\log q}{2\pi i} \int_\sigma^{\sigma+2\pi i/\log q} \mathcal{M}a(s)q^{-ns}\, ds, \tag{4.2}$$

independent of the complex number σ.

The *conjugate* of a is $a^*(x) = \bar{a}(1/x)$. Via the Mellin transform, this corresponds to $f^*(s) = \bar{f}(-\bar{s})$.

The *convolution product* of two sequences a and b in $\mathbb{A}^*/\ker|\cdot|$ is

$$(a * b)(\mathsf{t}^n) = \sum_{k=-\infty}^{\infty} a(\mathsf{t}^k)b(\mathsf{t}^{n-k}).$$

It corresponds to the product of functions, $\mathcal{M}(a*b)(s) = \mathcal{M}a(s)\mathcal{M}b(s)$.

Defining the norm by

$$\|a\|_*^2 = \sum_{n=-\infty}^{\infty} |a(\mathsf{t}^n)|^2,$$

we have $(a^* * a)(1) = \|a\|_*^2$.

4.2 Zeros of $\Lambda_{\mathcal{C}}$

For the function $\Lambda_{\mathcal{C}}(s)$ of Definition 3.4.1, the Riemann hypothesis says that all zeros of $\Lambda_{\mathcal{C}}$ are purely imaginary. This function is real-valued on $\mathbb{R}$, hence if ρ is a zero of $\Lambda_{\mathcal{C}}$, then $-\rho$, $\bar{\rho}$, and $-\bar{\rho}$ are zeros as well, all of the same multiplicity (see Figure 4.1). Depending on the location of ρ, this results in four or two different zeros of $\zeta_{\mathcal{C}}$, or just the zero ρ itself, as we now explain.

A complex number ρ is purely imaginary if and only if $-\bar{\rho} = \rho$. In that case, ρ and $-\rho$ are zeros of $\Lambda_{\mathcal{C}}$, $\bar{\rho}$ and $-\bar{\rho}$ being equal to $-\rho$ and ρ,

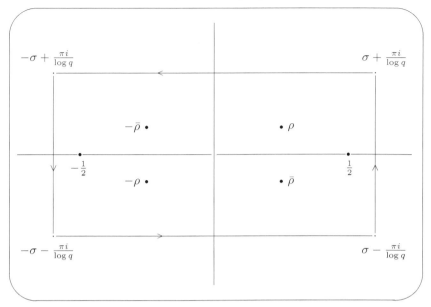

Figure 4.1 Four zeros of $\Lambda_{\mathcal{C}}$.

respectively. Also, if ρ is a real zero (or has imaginary part $\pi i/\log q$), then ρ and $-\rho$ are zeros of $\Lambda_{\mathcal{C}}$, $\bar{\rho}$ and $-\bar{\rho}$ now being equal to ρ and $-\rho$, respectively (modulo $2\pi i/\log q$ in the second case, by the periodicity of $\Lambda_{\mathcal{C}}$). Finally, if $\rho = 0$ or $\rho = \pi i/\log q$ is a zero of $\Lambda_{\mathcal{C}}$ then we cannot conclude that some other point is a zero as well. However, since $L_{\mathcal{C}}$ is an even analytic function, the multiplicity of $L_{\mathcal{C}}$ at 0 is even. Similarly, the multiplicity at $\pi i/\log q$ is even as well.

Lemma 4.2.1 *The Riemann hypothesis for $\mathcal{C}$ is equivalent to*

$$\sum_{\Lambda_{\mathcal{C}}(\rho)=0} f^*(\rho)f(\rho) \geq 0,$$

for every periodic function f, holomorphic in the strip $|\operatorname{Re} s| \leq 1/2$, with period $2\pi i/\log q$. Here, we can restrict to trigonometric polynomials, and even to such polynomials with $f(1/2) = f(-1/2) = 0$.

Proof For each purely imaginary zero ρ, we have $-\bar{\rho} = \rho$. Hence

$$f^*(\rho)f(\rho) = \bar{f}(-\bar{\rho})f(\rho) = |f(\rho)|^2 \geq 0.$$

Therefore, if all zeros are purely imaginary, then $\sum_{\rho} f^*(\rho)f(\rho) \geq 0$.

Conversely, if a zero ρ_0 is not purely imaginary, then $\rho_0 \neq -\bar{\rho}_0$. Hence we can construct a polynomial such that $f(\rho_0) = 1$, $f(-\bar{\rho}_0) = -1$, and f vanishes at all the other zeros that occur in the sum. Then $\sum_\rho f^*(\rho)f(\rho)$ is equal to minus twice the multiplicity of ρ_0, hence the sum is negative.

If it is known that x is not a zero of Λ_C, then this argument works for the subset of trigonometric polynomials that vanish at x. In particular, since Λ_C has a pole at $x = 1/2$ and at $x = -1/2$, we can restrict to polynomials that vanish at these points. $\qquad\square$

4.3 Explicit formula

Let f be a holomorphic periodic function in a neighborhood of the closed region enclosed by the contour R of Figure 4.1. By the residue theorem, we obtain for $\sigma > 1/2$ that

$$\frac{1}{2\pi i} \int_R f(s) \frac{\Lambda_C'(s)}{\Lambda_C(s)} \, ds = -f(1/2) - f(-1/2) + \sum_{\Lambda_C(\rho)=0} f(\rho),$$

where the sum is restricted to the zeros of Λ_C in one period.

Since f and Λ_C are periodic with period $2\pi i / \log q$, the integrals over the upper and lower side of R cancel each other. Further, Λ_C is even, so that the integral over the left-hand side equals

$$\frac{1}{2\pi i} \int_{\sigma - \frac{\pi i}{\log q}}^{\sigma + \frac{\pi i}{\log q}} f(-s) \frac{\Lambda_C'(s)}{\Lambda_C(s)} \, ds.$$

Thus we obtain for $\sigma > 1/2$,

$$-f(1/2) - f(-1/2) + \sum_{\Lambda_C(\rho)=0} f(\rho)$$

$$= \frac{1}{2\pi i} \int_{\sigma - \frac{\pi i}{\log q}}^{\sigma + \frac{\pi i}{\log q}} (f(s) + f(-s)) \frac{\Lambda_C'(s)}{\Lambda_C(s)} \, ds. \tag{4.3}$$

By formula (3.23) (where $q_v = q^{\deg v}$), we have for $\operatorname{Re} s > 1/2$,

$$\frac{1}{\log q} \frac{\Lambda_C'(s)}{\Lambda_C(s)} = g - 1 - \sum_v \deg v \sum_{n=1}^\infty q_v^{-n(s+1/2)}.$$

Let $f(s) = \mathcal{M}a(s)$ for a sequence a. Then we can apply the inversion formula (4.2) to obtain from (4.3),

$$-f(1/2) - f(-1/2) + \sum_{\Lambda_C(\rho)=0} f(\rho)$$
$$= (2g - 2)a(1) - \sum_v \deg v \sum_{n \neq 0} q_v^{-|n|/2} a(\pi_v^n),$$
(4.4)

where π_v denotes an idele of norm q_v^{-1}.

Remark 4.3.1 Formula (4.4) should be compared with the distributional explicit formulas in number theory [W6, p. 262]. For the trivial character, we can write this formula as follows:

$$-f(1/2) - f(-1/2) + \sum_{\rho} f(\rho)$$
$$= a(1) \log A - \sum_{v \nmid \infty} \log(Nv) \sum_{n \neq 0} (Nv)^{-|n|/2} a((Nv)^n)$$
$$- \sum_{v | \infty} \mathrm{PF} \int_0^\infty K_{v,0}(x) a(x) \, d^*x,$$

keeping some of the notation of Weil.

Defining the *Weil distribution*

$$W_v(a) = \sum_{n \neq 0} q_v^{-|n|/2} a(\pi_v^n), \qquad (4.5)$$

we have shown that

$$\sum_{\Lambda_C(\rho)=0} f(\rho) = f(1/2) + f(-1/2) + (2g - 2)a(1) - \sum_v \deg v \, W_v(a).$$
(4.6)

Applying this to $f^*(s)f(s) = \mathcal{M}(a^* * a)(s)$, we obtain

$$\sum_{\Lambda_C(\rho)=0} f^*(\rho)f(\rho) = 2 \operatorname{Re} f^*(1/2)f(1/2) + (2g - 2)\|a\|_*^2$$
$$- \sum_v \deg v \, W_v(a^* * a).$$
(4.7)

By Lemma 4.2.1, we have proved the following theorem:

Theorem 4.3.2 *The Riemann hypothesis is equivalent to*

$$(2g - 2)\|a\|_*^2 - \sum_v \deg v \, W_v(a^* * a) \geq 0$$

for all sequences a with $\mathcal{M}a(1/2) \cdot \mathcal{M}a(-1/2) = 0$. Moreover, it suffices to establish positivity for all finite sequences that satisfy this condition.

Example 4.3.3 By Example 3.4.3, the number V_d of valuations of degree d satisfies

$$\sum_{d|n} dV_d = q^n + 1 - \omega_1^n - \cdots - \omega_{2g}^n. \tag{4.8}$$

We can use this to give another proof of the explicit formula. Let

$$W_m(a) = \sum_{n \neq 0} q^{-m|n|/2} a(\mathfrak{t}^{mn})$$

be the Weil distribution for each valuation of degree m. Then the sum over the Weil distributions in the explicit formula reads

$$\sum_{m=1}^{\infty} m V_m W_m(a) = \sum_{m=1}^{\infty} m V_m \sum_{n \neq 0} q^{-m|n|/2} a(\mathfrak{t}^{mn})$$

$$= \sum_{m=1}^{\infty} \sum_{n \neq 0,\, m|n} m V_m q^{-|n|/2} a(\mathfrak{t}^n).$$

Interchanging the order of summation, we find in view of (4.8) that

$$\sum_{m=1}^{\infty} m V_m W_m(a) = \sum_{n \neq 0} q^{-|n|/2} a(\mathfrak{t}^n) \left(q^{|n|} + 1 - \omega_1^{|n|} - \cdots - \omega_{2g}^{|n|} \right).$$

For $f(s) = \mathcal{M}a(s)$, we find

$$\sum_{m=1}^{\infty} m V_m W_m(a) = (2g - 2)a(1) + f(1/2) + f(-1/2) - \sum_{\nu=1}^{2g} f(\rho_\nu).$$

Remark 4.3.4 In the next chapter, Section 5.4, we will see the following relationship with the number of points on $\mathcal{C}$:

$$-f(1/2) - f(-1/2) + \sum_{\Lambda_\mathcal{C}(\rho)=0} f(\rho)$$

$$= (2g - 2)a(1) - \sum_{n \neq 0} q^{-|n|/2} N_\mathcal{C}(|n|) a(\mathfrak{t}^n),$$

where $\mathfrak{t}$ denotes an idele of norm q and $N_\mathcal{C}(n)$ is the number of points on $\mathcal{C}$ defined over $\mathbb{F}_{q^n}$.

5

The Frobenius flow

A curve defined over a finite field $\mathbb{F}_q$ has points with coordinates in $\mathbb{F}_q^a$. The Frobenius automorphism acts on these points by raising each coordinate to the q-th power. We call this action the Frobenius flow on $\mathcal{C}$. The fixed points of this flow are the points of $\mathcal{C}(\mathbb{F}_q)$, the points on $\mathcal{C}$ defined over $\mathbb{F}_q$.

We show that the orbits of the Frobenius flow on a curve are in direct correspondence with the valuations of the field of functions on the curve. We first obtain this correspondence for the projective line, with function field $\mathbf{q} = \mathbb{F}_q(T)$, and then in general for any curve $\mathcal{C}$.

In Section 5.4, we present Bombieri's proof of the Riemann hypothesis for a curve over a finite field, after we have reformulated the Riemann hypothesis as an estimate on the number of points on $\mathcal{C}$.

Bombieri's proof involves Riemann–Roch in an essential way (but the full strength of Theorem 3.1.9 is not needed). In addition, it uses some combinatorics of functions in two variables on $\mathcal{C} \times \mathcal{C}$.

In Chapter 6, we discuss another approach to the Riemann hypothesis, which only involves the use of Riemann–Roch.

5.1 Heuristics for the Riemann hypothesis

We start this chapter with a discussion of the heuristics for the Riemann hypothesis as described in Connes' paper [Conn1]. The reader is referred to this paper for the analogy with Selberg's trace formula. See also Remark 6.3.5.

Recall the average restriction E. It restricts a function on the adeles to a function on the coarse idele class group. As we have seen in Exercise 3.4.2, the coefficients of the power series for $\zeta_{\mathcal{C}}$ are expressed in terms of this operator.

The coarse idele class group $Z = \mathbb{A}^*/\ker|\cdot|$ is isomorphic to $\mathbb{Z}$ via the degree $\mathbb{A}^* \to \mathbb{Z}$. Multiplication by an idele t of degree 1 induces a shift on this space,

$$Sf(a) = f(at).$$

For a function h on Z with finite support, we let S_h be the combined shift

$$S_h = \sum_{n \in Z} h(n) S^{\deg n}.$$

To turn S_h into a trace class operator, we compose it with a projection of finite rank. Given two positive numbers ω and θ, we denote by

$$Z_{-\omega..\theta}$$

the projection onto the space of functions supported in $-\omega \leq \deg a \leq \theta$, a space of dimension $\omega + \theta + 1$. In Section 6.1, we will compute the trace,

$$\mathrm{tr}\big(Z_{-\omega..\theta} S_h\big) = (\omega + \theta + 1)h(0).$$

We see that the shift on the coarse idele class group has no interaction.

We may require that $\mathcal{M}h(1/2) = \mathcal{M}h(-1/2) = 0$, so that $\mathcal{M}h(s)$ is a multiple of

$$d(s) = \big(1 - q^{1/2-s}\big)\big(1 - q^{1/2+s}\big).$$

The space of all such shifts inside $Z_{-\omega..\theta}$ has dimension $\omega + \theta - 1$. We denote that space by $P_{-\omega..\theta}$; it is the space of all Fourier series supported in $-\omega..\theta$ that are a multiple of $d(s)$. Then

$$\mathrm{tr}\big(P_{-\omega..\theta} S_h\big) = (\omega + \theta - 1)h(0) + o(1),$$

where $o(1)$ denotes a function that tends to 0 as $\omega, \theta \to \infty$.

Consider now the truncation on the adeles as follows. We choose two divisors Ω and Θ, respectively of degrees ω and θ,

$$\deg \Omega = \omega \quad \text{and} \quad \deg \Theta = \theta.$$

The truncation is in three steps: first choose a lift (possibly using an operator L), then cut off this lift on the adeles (possibly using an ultraviolet and an infrared truncation, C_θ and M_Ω, to be introduced in Chapter 6), and then apply E to restrict to the coarse ideles class group again,[1]

$$L^2(\mathbb{A}/\mathfrak{o}^*) \xrightarrow{C_\theta M_\Omega} L^2(\mathbb{A}/\mathfrak{o}^*)$$

$$L \Big\uparrow \qquad\qquad \Big\downarrow E$$

$$l^2(Z) \xrightarrow{\quad S \quad} l^2(Z) \cdots\cdots\xrightarrow{\quad B_{\Omega,\theta} \quad} l^2(Z)$$

Remark 5.1.1 (i) The notation Ω for a divisor of (large) degree ω is intended to remind the reader of the large wave lengths of the infrared truncation, and Θ (of degree θ) should remind the reader of the high frequency waves of the ultraviolet truncation. See Sections 6.3–6.7.

(ii) Lemma 3.1.6 says that the Fourier transform on the additive space of adeles interchanges the small and large scale behaviour. In Chapter 6, M_Ω will be defined as a truncation operator that restricts the support of a function, and C_θ will smooth out the small oscillations of a function. On the space of frequencies (the coarse idele class group), this is simply a truncation to a finite band width.

There are two requirements on both the infrared and ultraviolet truncations on the adeles. The combination should project onto functions supported in

$$\pi^{-\Omega}\mathfrak{o}$$

that are locally constant at distances in

$$\pi^{-\mathcal{K}+\Theta}\mathfrak{o}.$$

Moreover, every function in the image must vanish at zero together with its Fourier transform, in view of (3.11).

The first two requirements imply that

$$\Omega \geq \mathcal{K} - \Theta,$$

and if there is a nontrivial function satisfying $f(0) = \mathcal{F}f(0) = 0$, then

$$\omega + \theta \geq \deg \mathcal{K} + 2 = 2g.$$

Finally, it is necessary that the projections M_Ω and C_θ commute.

With these requirements, any function in the image vanishes on ideles of degree less than $-\omega$, since it lies in the image of M_Ω:

$$Ef(a) = 0 \quad \text{for } f \in \text{Im}(C_\theta M_\Omega), \ \deg a < -\omega.$$

----

[1] We have no definite idea how to define the operator L, as indicated by the dotted arrow. We discuss one possible approach in Chapter 6.

Since f is locally constant at distances in $\pi^{-\mathcal{K}+\Theta}\mathfrak{o}$, by Lemma 3.1.6, its Fourier transform is supported in $\pi^{-\Theta}\mathfrak{o}$. By (3.11), since f and $\mathcal{F}f$ vanish at 0, we have

$$Ef(a) = E(\mathcal{F}f)(1/a),$$

and hence $Ef(a)$ vanishes for $\deg 1/a < -\theta$. It follows that Ef is supported in $-\omega \le \deg a \le \theta$. We see that the projection $B_{\Omega,\theta} = EC_\theta M_\Omega L$ is subordinate to $P_{-\omega..\theta}$,

$$B_{\Omega,\theta} \le P_{-\omega..\theta} \tag{5.1}$$

in the sense of positive operators.

Our goal is to obtain the explicit formula,

$$\mathrm{tr}(B_{\Omega,\theta}S_h) = (\omega + \theta + 1 - 2g)h(0) + \sum_v \deg v \, W_v(h).$$

With the right choice of M_Ω and C_θ, this may be possible to achieve. Comparing with the trace of the truncation with $Z_{-\omega..\theta}$, we would obtain, for $h = a^* * a$ and $\mathcal{M}h(1/2) = \mathcal{M}h(-1/2) = 0$,

$$(\omega + \theta + 1 - 2g)h(0) + \sum_v \deg v \, W_v(h) \le (\omega + \theta - 1)h(0) + o(1).$$

Since the contribution from ω and θ cancels, the $o(1)$ function would actually vanish. Thus we would obtain Weil positivity, Theorem 4.3.2, and hence the Riemann hypothesis for $\mathcal{C}$.

So the rather simple vanishing properties of the restriction E would imply the Riemann hypothesis. Since the restriction is intimately connected with the theorem of Riemann–Roch, we should ask ourselves if it is reasonable to expect that Riemann–Roch is strong enough to imply the Riemann hypothesis. In the present chapter, we study Bombieri's proof, which uses Riemann–Roch and some geometry on $\mathcal{C} \times \mathcal{C}$. Even though the geometric facts that we use in the proof are not very deep, the mere fact that we use a two-dimensional space makes the translation to $\mathrm{spec}\,\mathbb{Z}$ highly nontrivial, if not impossible. On the other hand, the translation of Riemann–Roch is not a problem, and can already be found in Tate's thesis [Ta].

5.2 Orbits of Frobenius

Let $\mathcal{C}$ be a geometrically irreducible curve[2] with function field K defined over a finite field $\mathbb{F}_q$ of characteristic p. We assume that $\mathbb{F}_q$ is algebraically closed in K, so that $\mathbb{F}_q$ is the field of constants of K. For a valuation v of K and a function f on $\mathcal{C}$, we say that f is *regular* at v if $v(f) \geq 0$, that v is a *zero* of f if $v(f) > 0$, and that f has a *pole* at v if $v(f) < 0$. Recall from Section 1.2.2 that the completion K_v contains the local ring $\mathfrak{o}_v$ of germs of functions regular at v, with prime ideal $\pi_v \mathfrak{o}_v$ of germs that vanish at v.

The field of residue classes is $K(v) = \mathfrak{o}_v / \pi_v \mathfrak{o}_v$, a finite extension of $\mathbb{F}_q$. It is the field of values at v of germs of regular functions. It is geometrically meaningful to add ∞ to this set of values to obtain $\mathbb{P}^1(K(v))$ as the set of values at v of functions in K. Thus we can write $f(v) = \infty \in \mathbb{P}^1(K(v))$ if f has a pole at v.

The *degree* of v is

$$\deg v = [K(v) : \mathbb{F}_q]. \tag{5.2}$$

We will see below that the geometric meaning of $\deg v$ is the number of points on $\mathcal{C}$ corresponding to v.

In order to describe K more explicitly, we choose a nonconstant function T in K such that K is a finite extension of $\mathbb{F}_q(T)$, which we assume to be separable; i.e., T is not a p-th power in K.[3] Then

$$K = \mathbb{F}_q(T)[X]/(m),$$

where $m(T, X) = 0$ is an equation for $\mathcal{C}$. This curve may be singular. We will only study its regularization, using a nonsingular model for $\mathcal{C}$.

5.2.1 The projective line

The simplest function field is $\mathbf{q} = \mathbb{F}_q(T)$, the extension of $\mathbb{F}_q$ generated by one transcendental element. By Theorem 1.3.2, the valuations of $\mathbf{q}$ are the finite valuations v_P, for each irreducible polynomial P, and the valuation at infinity. By Remark 1.3.4, this last valuation is associated

---

[2] For example, the curve $X^2 + X = T^2 + T + 1$, which is irreducible over $\mathbb{F}_2$, has two components $X = T + \alpha$ and $X = T + \alpha + 1$ over $\mathbb{F}_4 = \mathbb{F}_2[\alpha]$. Geometrically irreducible means that the equation for the curve is irreducible over the algebraic closure $\mathbb{F}_q^a$.

[3] Otherwise, $\sqrt[p]{T}$ lies in K and K is a finite extension of $\mathbb{F}_q(\sqrt[p]{T})$. If this is still not a separable extension, we continue in this manner to obtain a separable extension after finitely many steps. See Section 1.2.4.

with $P = 1/T$. We normalize the valuations by

$$v_{1/T}(T) = -1 \quad \text{and} \quad v_P(P) = 1.$$

To compute $v_P(f)$ for a polynomial f and a finite valuation v_P, we write f in base P:

$$f(T) = f_0(T) + f_1(T)P + \cdots + f_n(T)P^n + \cdots + f_m(T)P^m, \quad (5.3)$$

where the coefficients $f_0, f_1, \ldots$ are polynomials in T of degree less than the degree of P. If $f_0 = f_1 = \cdots = f_{n-1} = 0$ and f_n does not vanish identically, then $v_P(f) = n$.

The class of f in $\mathbf{q}(v_P)$ is uniquely represented by $f_0(T)$, since $P = 0$ in $\mathbf{q}(v_P)$. Hence $[\mathbf{q}(v_P) : \mathbb{F}_q] = \deg P$. By our definition of the degree of a valuation (5.2), we obtain

$$\deg v_P = \deg P.$$

We first consider the case when P is linear, say $P(T) = T - t$ for some value $t \in \mathbb{F}_q$. Thus the coefficients f_i in (5.3) are constants in $\mathbb{F}_q$. Since $P(t) = 0$, the class of f in $\mathbf{q}(v_P)$ is represented by $f_0 = f(t)$, the value of f at t. For a rational function $f = (T - t)^k \alpha(T)/\beta(T)$, with polynomials $\alpha(T)$ and $\beta(T)$ as in (5.3) and $\alpha_0, \beta_0 \neq 0$, we find

$$f(t) = 0 \quad \text{if } k > 0,$$
$$f(t) \in \mathbb{F}_q^* \quad \text{if } k = 0, \text{ and}$$
$$f(t) = \infty \quad \text{if } k < 0.$$

Moreover, $k = v_P(f)$ equals the order of vanishing of f at t. Thus v_P corresponds to the point $(1 : t) \in \mathbb{P}^1(\mathbb{F}_q^a)$, which is the point where the polynomial P vanishes. The class of f in $\mathbf{q}(v_P)$ equals the value of f at this point.

Exercise 5.2.1 Show that the value of a polynomial $f \in \mathbb{C}[T]$ at the complex number t can be found as the residue of $f(T)$ when divided by $T - t$.

Since the valuation $v_{1/T}$ also has degree one, we expect a similar description for this valuation. Indeed, every function can be written as a quotient of polynomials in $1/T$. In projective coordinates $(S : T)$, a polynomial in $1/T$ is an expression

$$f(S : T) = f_0 + f_1 \frac{S}{T} + \cdots + f_n \left(\frac{S}{T}\right)^n + \cdots + f_m \left(\frac{S}{T}\right)^m, \quad (5.4)$$

where $f_i \in \mathbb{F}_q$. We find that $f_0 = f(0 : 1)$ is the value of f at the point at infinity on $\mathbb{P}^1$. For a quotient $f(S : T) = (S/T)^k \alpha/\beta$ of such polynomials, with α and β as in (5.4) and $\alpha_0, \beta_0 \neq 0$, we find

$$f(0 : 1) = 0 \quad \text{if } k > 0,$$
$$f(0 : 1) \in \mathbb{F}_q^* \quad \text{if } k = 0, \text{ and}$$
$$f(0 : 1) = \infty \quad \text{if } k < 0.$$

Moreover, $k = v_{1/T}(f)$ equals the order of vanishing of f at $(0 : 1)$. Thus the valuation $v_{1/T}$ corresponds to the point at infinity on the projective line, which is the point where the function $1/T$ vanishes, and the class of f in $\mathbf{q}(v_{1/T})$ equals the value of f at this point.

We now consider a valuation of degree more than one, which is necessarily a finite valuation. Since we work over a finite field, adjoining one root x of P to $\mathbb{F}_q$ gives the splitting field $\mathbb{F}_q[x] = \mathbf{q}(v_P)$ of P over $\mathbb{F}_q$. The conjugates of x are obtained by applying the Frobenius automorphism

$$\phi_q \colon x \longmapsto x^q,$$

which generates the Galois group of $\mathbb{F}_q^a$ over $\mathbb{F}_q$. We adjoin x to $\mathbf{q}$ to obtain the constant field extension $E = \mathbf{q}[X]/(P)$. By Lemma 1.4.1, to see how v_P extends to E, we need to consider the factorization of $P(X)$ over the completion $\mathbf{q}_P$. The ring of integers of $\mathbf{q}_P$ consists of (infinite) power series (5.3). It turns out that $\mathbf{q}_P$ contains a root of P, and that we can find this root by an application of Hensel's lemma (Newton's algorithm).

Construction 5.2.2 (Hensel's Lemma) *Let K be a complete field with a discrete valuation v. Let*

$$P(X) = p_0 + p_1 X + \cdots + p_d X^d$$

be a polynomial over K with coefficients $p_i \in \mathfrak{o}_v$, and let $a \in \mathfrak{o}_v$ be such that $v(P(a)) > 0$ and $v(P'(a)) = 0$. Construct a root of P in K.

Algorithm Let π be a uniformizer for v. The number a will be the first digit of a root $x = a + a_1\pi + a_2\pi^2 + \cdots$.

The first approximation of a root is $x_0 = a$. Then $P(x_0) = b\pi$ for some $b \in \mathfrak{o}_v$. Suppose now that we have computed

$$x_n = a + a_1\pi + \cdots + a_n\pi^n \qquad (n \geq 0),$$

with $a_i \in \mathfrak{o}_v$ such that $P(x_n)$ is divisible by π^{n+1}. Then $P\left(x_n + a_{n+1}\pi^{n+1}\right)$ can be expanded,

$$P\left(x_n + a_{n+1}\pi^{n+1}\right) = \sum_{k=0}^{d} p_k \sum_{i=0}^{k} \binom{k}{i} x_n^{k-i} a_{n+1}^i \pi^{(n+1)i}. \qquad (5.5)$$

We consider the remainder of (5.5) modulo π^{n+2}. For $i = 0$, the sum over k gives $P(x_n)$, which has a remainder $b\pi^{n+1}$ for some $b \in \mathfrak{o}_v$ by assumption. For $i = 1$, the sum over k is $P'(x_n)a_{n+1}\pi^{n+1}$, which has remainder $P'(a)a_{n+1}\pi^{n+1}$ since $P'(x_n) = P'(a) \bmod \pi$. And for $i \geq 2$, each term vanishes modulo π^{n+2}. Thus we find the next digit

$$a_{n+1} = -b/P'(a).$$

Since $1/P'(a) \in \mathfrak{o}_v$, we have $a_{n+1} \in \mathfrak{o}_v$. It follows that

$$x_{n+1} = x_n + a_{n+1}\pi^{n+1} = x_n - \frac{P(x_n)}{P'(x_n)} \bmod \pi^{n+2},$$

so that the series for x converges in K.

At each stage of the construction, $P(x) = P(x_n) = 0 \bmod \pi^{n+1}$. It follows that $P(x) = 0$. $\qquad \square$

The foregoing algorithm adds one correct new digit of the expansion of x at each turn. By the following exercise, it is useful to keep a much larger number of digits of the expansion of $x_n - P(x_n)/P'(x_n)$.

Exercise 5.2.3   Show that the recursive formula

$$x_{n+1} = x_n - \frac{P(x_n)}{P'(x_n)}$$

gives an algorithm that doubles the number of correct digits with each turn.

For the next exercise, see also [Bak, Theorem 12.16].

Exercise 5.2.4   Describe how the condition $v(P'(a)) = 0$ can be weakened if the condition $v(P(a)) > 0$ is strengthened.

Recall that P is an irreducible polynomial with coefficients in $\mathbb{F}_q$ and $E = \mathbf{q}[X]/(P)$. Since $P(T) = 0 \bmod P$, the element $T \in \mathbf{q}$ approximates a root of P. Moreover, $P'(T) \neq 0 \bmod P$. By Hensel's lemma, we find a root $x \in \mathbf{q}_P$ of P. Then $x, \phi_q(x), \ldots, \phi_q^{d-1}(x)$ are all the roots of P, where $d = \deg P$. Thus P splits into linear factors over $\mathbf{q}_P$. By Lemma 1.4.1, there are $\deg P$ extensions of v_P to E. Since the factors of P are linear, each of these valuations of E corresponds to a point on the projective line over the finite field $\mathbf{q}(v_P)$, as we have seen above for valuations of degree one. Thus v_P itself corresponds to the *orbit of Frobenius*

$$\{(1 : \phi_q^n(x)) : n \doteq 0, \ldots, \deg P - 1\} = \{(1 : x) : P(x) = 0\},$$

on the projective line over $\mathbb{F}_q^a$. Clearly, v_P is the only valuation that gives this set of points on $\mathbb{P}^1(\mathbb{F}_q^a)$.

For a rational function $f = P^k g$, where g is a rational function without a factor P in its numerator or denominator, we find, for each point x where P vanishes,

$$f(x) = 0 \quad \text{if } k > 0,$$
$$f(x) = g \bmod P \quad \text{if } k = 0, \text{ and}$$
$$f(x) = \infty \quad \text{if } k < 0.$$

Moreover, $k = v_P(f)$ equals the order of vanishing of f at any one of these points. Note that since f is defined over $\mathbb{F}_q$, these orders all coincide. Finally, the class of f in $\mathbf{q}(v_P)$ is $f(T) + (P)$, which is the value of f at the root $T + (P)$ of P in $\mathbf{q}(v_P) = \mathbb{F}_q[T]/(P)$. This field is not a subfield of $\mathbf{q}$, since some of its elements, for example $T + (P)$, are only represented by nonconstant functions in $\mathbf{q}$. But it becomes a subfield after extending the field of constants to $\mathbf{q}(v_P)$.

We have proved the following theorem:

Theorem 5.2.5 *There exists a correspondence between valuations of* $\mathbf{q}$ *and orbits of the Frobenius flow* ϕ_q *on* $\mathbb{P}^1(\mathbb{F}_q^a)$. *Under this correspondence, a valuation* v *corresponds to the orbit*

$$\{x \in \mathbb{P}^1(\mathbb{F}_q^a) : f(x) = 0 \text{ for all } f \text{ with } v(f) > 0\},$$

and an orbit $\{\phi_q^n(x) : n \in \mathbb{Z}\}$ *corresponds to the valuation* $v_x(f)$. *The degree of this valuation is* $\deg v = |\{\phi_q^n(x) : n \in \mathbb{Z}\}|$, *the number of points in the orbit, which is also the dimension of* $\mathbf{q}(v)$ *over* $\mathbb{F}_q$.

We have now three ways to describe a valuation. Algebraically, we have

$$v_P \text{ and } v_{1/T},$$

giving the number of factors P of a function, or minus its degree. Here, P is an irreducible polynomial.

In terms of the geometry of points on the curve, we have

$$v_x, \text{ for each } x \in \mathbb{P}^1(\mathbb{F}_q^a),$$

giving the order of vanishing of a function at the point x.

Finally, we have a dynamical interpretation of the valuations, in terms of the orbits of the Frobenius flow,

$$v_{\{\phi_q^n(x):\, n\in\mathbb{Z}\}}, \text{ for each orbit of } x \in \mathbb{P}^1(\mathbb{F}_q^a),$$

since the valuations v_x for the different points in an orbit of the Frobenius flow on $\mathbb{P}^1$ coincide. This orbit is finite and equal to the divisor of zeros of the corresponding irreducible polynomial P.

5.2.2 Example: elliptic curves

We explain a well-known method to obtain a Weierstrass model for an elliptic curve.

By Example 3.1.12, if K is a function field of genus one, then K has a valuation of degree one. We denote this valuation by ∞ and consider it the point at infinity. Then $l(n\infty) = n$ for $n \geq 1$. Applying this to $n = 1$, 2 and 3, we find the constant function $1 \in L(\infty)$, a function

$$t \in L(2\infty)$$

with a pole of order two at infinity, and a function

$$x \in L(3\infty)$$

with pole divisor $3(\infty)$. Then $(1, t, x, t^2, tx)$ is an $\mathbb{F}_q$-basis for $L(5\infty)$, where the constant function 1 has no pole at ∞, and the order of the pole at ∞ of the other functions is two, three, four, and five, respectively.

We find two functions with a pole of order six at infinity, t^3 and x^2. These are not independent, hence there exist constants a_0, a_1, a_2, a_3, a_4 and a_6 in the residue class field $K(\infty) = \mathbb{F}_q$ such that

$$x^2 = a_0 t^3 + a_1 xt + a_2 t^2 + a_3 x + a_4 t + a_6, \tag{5.6}$$

and this gives a nonsingular equation for $\mathcal{C}$.

Every function with a pole at ∞ alone can be written as a polynomial in t and x. Indeed, if the pole divisor of f is $n(\infty)$, then $n \geq 2$ (otherwise f is an isomorphism of $\mathcal{C}$ with $\mathbb{P}^1$). If $n \geq 6$ then we can find k and l such that $2k + 3l = n$ (see Lemma 5.2.6 below). Then $ft^{-k}x^{-l}$ has no pole at ∞, hence a value $c \in \mathbb{F}_q$. It follows that $f - ct^k x^l$ has a pole of lower order at ∞. Continuing this way, we arrive at a function with a pole of order at most five, for which t, x, t^2 or tx can be used to cancel the pole, eventually arriving at a constant function.

That the Weierstrass equation (5.6) is nonsingular follows from the fact that $\mathbb{F}_q[t, x]$ contains a local coordinate for every finite valuation, as we show in the next section. In other words, the criterion for regularity that we will use is that $\mathfrak{m}_v/\mathfrak{m}_v^2 \cong K(v)$ for every valuation, where $\mathfrak{m}_v$ is the maximal ideal of $\mathfrak{o}_v$.

5.2.3 The curve $\mathcal{C}$

We consider now an arbitrary curve of genus g with function field K. We assume that K has a valuation of degree 1, and denote this valuation by ∞. We can find two functions t and x with only a pole at ∞ and such that their orders at ∞ are relatively prime: since $l((2g-1)\infty) = g$ and $l(2g\infty) = g+1$, there exists a function t with a pole of order $2g$ at ∞. Similarly, adding one more (∞) to the divisor, there exists a function x with a pole of order $2g + 1$. Note that $\gcd(2g, 2g+1) = 1$.

Lemma 5.2.6 *Let a and b be positive integers with $\gcd(a, b) = 1$. Then every integer $\geq ab$ can be written as a linear combination $ka + lb$ with positive coefficients.*

Proof Since $\gcd(a, b) = 1$, let p and q be integers such that $pa + qb = 1$. We obtain $pna + qnb = n$. However, pn and qn will in general not be positive. Adjusting the coefficients by $(pn - mb)a + (qn + ma)b = n$, we want to find m such that $pn - mb \geq 0$ and $qn + ma \geq 0$, or

$$-\frac{q}{a}n \leq m \leq \frac{p}{b}n.$$

This interval has length $n/(ab)$, and for $n \geq ab$ it contains an integer. $\quad\square$

Lemma 5.2.7 *Let t and x be two functions with only a pole at infinity, of degrees a and b, respectively, such that $\gcd(a, b) = 1$. Then there exist finitely many functions $y_1, y_2, \ldots, y_m$ such that every function with only a pole at ∞ is a polynomial in $t, x, y_1, \ldots, y_m$.*

Proof By the previous lemma, if a function f has a pole of order at least ab at infinity, then there exist a constant $c \in \mathbb{F}_q$ and exponents k and l such that $f - ct^k x^l$ has a pole of lower order at ∞. Continuing in this fashion, we find a polynomial $P(T, X)$ with coefficients in $\mathbb{F}_q$ such that $f - P(t, x)$ has a pole of order at most $ab - 1$ at infinity. This pole can be cancelled using a basis $y_1, y_2, \ldots, y_m$ of $L((ab-1)\infty)$. We thus arrive at a function without poles that vanishes at ∞. Such a function vanishes identically. $\quad\square$

Given functions t, x, and y_i as in the lemma, consider the homomorphism

$$\mathbb{F}_q[T, X, Y_1, \ldots, Y_m] \longrightarrow K, \qquad T \mapsto t, \ X \mapsto x, \text{ and } Y_i \mapsto y_i.$$

By the lemma, this homomorphism is surjective onto the ring $K_{\{\infty\}}$ of functions that have a pole at ∞ alone (see (3.30)). Since the polynomial ring is Noetherian, the kernel is finitely generated, by $\{F_i\}$, say, and this gives equations for $\mathcal{C}$ in $m + 2$-dimensional affine space.

Lemma 5.2.8 *For every valuation $v \neq \infty$, there exists a function f with a pole at ∞ alone and such that $v(f) = 1$.*

Proof Let D be the divisor $(2 \deg v + 2g - 1)(\infty) - 2(v)$. Since its degree is $2g - 1$, the divisor $\mathcal{K} - D$ has a negative degree, hence $l(D) = g$. By the same reasoning, $l(D + (v)) = g + \deg v$. It follows that there are $\deg v$ functions in $L(D + (v))$ with a zero of order 1 at v. Any one of these functions satisfies the requirement. □

With this lemma, we see that $\mathbb{F}_q[t, x, y_1, \ldots, y_m]$ contains a uniformizer for every valuation. It follows that the curve given by the equations $\{F_i\}$ has no singularities, as illustrated by Lemma 5.2.10 and Examples 5.2.9 and 5.2.11 below.

Example 5.2.9 The curve $Y^2 = X^3$ is singular at $(X, Y) = (0, 0)$ (see the cusp singularity in Figure 5.1). The maximal ideal at this point is

$$\mathfrak{m} = (X, Y),$$

and one computes that the quotient

$$\mathfrak{m}/\mathfrak{m}^2 = X\mathbb{F}_q + Y\mathbb{F}_q$$

is two-dimensional over $\mathbb{F}_q$. This is explained by the parametrization $X = T^2$, $Y = T^3$, which gives

$$\mathfrak{m} = (T^2) \quad \text{and} \quad \mathfrak{m}/\mathfrak{m}^2 = T^2\mathbb{F}_q + T^3\mathbb{F}_q.$$

The ring $\mathbb{F}_q[X, Y]$ does not contain a function with a simple zero at $(0, 0)$, but the function $T = Y/X$ in the larger ring $\mathbb{F}_q[T]$ has a simple zero at this point.

A double point such as in the next example does not occur in our model for $\mathcal{C}$ either, because functions in $K_{\{\infty\}}$ separate points:

Lemma 5.2.10 *Given two valuations v, $w \neq \infty$, there exists a function f with a pole at ∞ alone and such that $v(f) \geq 1$ and $w(f) = 0$.*

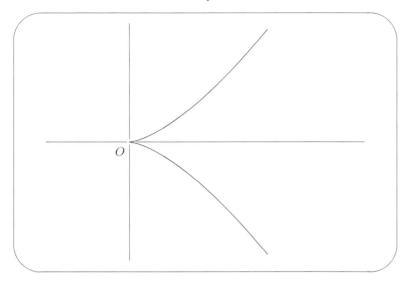

Figure 5.1 The curve $Y^2 = X^3$ has a cusp singularity.

Proof Let D be the divisor $(\deg v + \deg w + 2g - 1)(\infty) - (v) - (w)$. Clearly, $l(D) = g$ and $l(D + (w)) = g + \deg w$. Hence there exists a function with a pole at ∞ alone and a zero at v (i.e., $v(f) \geq 1$), that does not lie in $L(D)$, i.e., $w(f) = 0$. □

Example 5.2.11 The curve $Y^2 = X^3 + X^2$ of Figure 5.2 is singular at $(0, 0)$. The maximal ideal is

$$\mathfrak{m} = (X, Y),$$

and one computes that the quotient

$$\mathfrak{m}/\mathfrak{m}^2 = X\mathbb{F}_q + Y\mathbb{F}_q$$

is two-dimensional. The parametrization $X = T^2 - 1$, $Y = T(T^2 - 1)$ gives

$$\mathfrak{m} = (T^2 - 1) = (T - 1)(T + 1),$$

the product of the two maximal ideals in $\mathbb{F}_q[T]$ that correspond to v_{-1} and v_1. These two points are not separated in $\mathbb{F}_q[X, Y]$.

There is a one-to-one correspondence between finite valuations and orbits of points that satisfy the equations F_i. Given a point

$$p = (t(p), x(p), y_j(p))$$

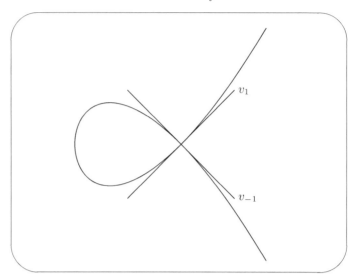

Figure 5.2 The curve $Y^2 = X^3 + X^2$ has a double point.

with coordinates in $\mathbb{F}_q^a$ such that $F_i(t(p), x(p), y_j(p)) = 0$, we obtain the function e_p of evaluation at p,

$$e_p \colon K_{\{\infty\}} \longrightarrow \mathbb{F}_q^a, \qquad f = F(t, x, y_j) \longmapsto e_p(f) = F(t(p), x(p), y_j(p)),$$

where F exists by Lemma 5.2.7.

The choice of F is not unique: it is determined up to elements of the ideal generated by F_i, the equations for $\mathcal{C}$. But since each F_i vanishes at p, the function e_p is well-defined.

The kernel of e_p is a maximal ideal that determines the valuation associated with p, and any other point in the orbit $\phi_q^n(p)$ determines the same valuation.

Conversely, a valuation determines the class $(t(v), x(v), y_j(v))$ of t, x, and y_j in $K(v)$. Since

$$F_i(t(v), x(v), y_j(v)) = F_i(t, x, y_j) \bmod \pi_v \mathfrak{o}_v,$$

this point satisfies the equations of $\mathcal{C}$.

Theorem 5.2.12 *There exists a correspondence between valuations of K and orbits of the Frobenius flow ϕ_q on $\mathcal{C}(\mathbb{F}_q^a)$. Under this correspondence, a valuation v corresponds to the orbit*

$$\{x \in \mathcal{C}(\mathbb{F}_q^a) \colon f(x) = 0 \ \text{for all} \ f \ \text{with} \ v(f) > 0\},$$

and an orbit $\{\phi_q^n(x)\colon n \in \mathbb{Z}\}$ corresponds to the valuation v_x. The degree of the valuation is $\deg v = |\{\phi_q^n(x)\colon n \in \mathbb{Z}\}|$, the number of points in the orbit, which is also the dimension of $K(v)$ over $\mathbb{F}_q$.

Algebraically, every valuation of K is obtained as an extension of a valuation of $\mathbf{q}$. The geometric point of view is no different than for $\mathbb{P}^1$: every point x in $\mathcal{C}(\mathbb{F}_q^a)$ gives a valuation v_x, defined as the order of vanishing of a function at x. And dynamically, there are $\deg(v_x)$ other points in an orbit of Frobenius that give the same valuation associated with x.

5.3 Galois covers

A cover $\mathcal{C}' \twoheadrightarrow \mathcal{C}$ corresponds to an extension L/K of function fields. A K-automorphism of L corresponds to an action $\mathcal{C}' \to \mathcal{C}'$, and this action is always algebraic, as is shown in the following proposition.

Proposition 5.3.1 *Let K be the function field of a curve $\mathcal{C}$, and let $\mathcal{C}'$ be a cover of $\mathcal{C}$ with function field L with a K-automorphism σ. Then we have an induced algebraic action of σ on $\mathcal{C}'$.*

Proof Write L as $K[X]/(m)$ for some function $x = X + (m) \in L$ with defining polynomial m, and let $\sigma \in \mathrm{Gal}(L/K)$. Since $\sigma(x) \in L$, we can find a polynomial f with coefficients in K such that $\sigma(x) = f(X) + (m)$. Thus the action of σ is algebraic, induced by $X \mapsto f(X)$.

On the points of $\mathcal{C}'$, the action of σ is obtained as follows: we need a model for $\mathcal{C}'$, given with coordinates (t', x', y_i'), say, as in the previous section. Then t' can be written as a rational function of X modulo m. Replacing X by $f(X)$ gives the action of σ on t'. But $X + (m)$ itself is a rational expression in t', x', and y_i', and we thus find the image of t' as a rational expression in the coordinates of $\mathcal{C}'$. Doing the same for the other coordinates, we obtain a complete description of the action of σ. $\quad\square$

We know from Galois theory that the field generated by the roots of a separable polynomial is a Galois extension, and that, conversely, every finite Galois extension is the splitting field of some polynomial. It follows that if L/K is Galois, say L is the splitting field of the polynomial m, and w is a valuation on L restricting to v on K, then L_w/K_v is also a Galois extension, since L_w is the splitting field of m over K_v. This gives one way to understand the following proposition, but we give a more concrete proof.

Proposition 5.3.2 *Let $L = K[X]/(m)$ be a Galois extension, and let v be a valuation of K. Let w and w' be two extensions of v to L. Then L_w and $L_{w'}$ are isomorphic.*

Proof Let $x \in L$ be a root of m_w, and $x' \in L$ of $m_{w'}$, where m_w and $m_{w'}$ are the factors of m corresponding to w and w' (see Notation 1.4.2). Since x and x' are both roots of m and L/K is Galois, there exists an automorphism σ of L such that $\sigma x = x'$. The map $K_v[X] \to K_v[X]/(m_{w'})$ given by $X \mapsto \sigma x$ has kernel (m_w). Hence σ gives an isomorphism $L_w \to L_{w'}$.   □

Let $L = K[X]/(m)$ be a Galois extension with Galois group $\mathrm{Gal}(L/K)$. Each automorphism of L is determined by the image of X. Let w be a valuation of L extending v. Since an automorphism of L_w over K_v is again determined by the image of X, the group $\mathrm{Gal}(L_w/K_v)$ can be identified with a subgroup of $\mathrm{Gal}(L/K)$.

Definition 5.3.3 The group

$$Z_{w/v} = \{\sigma \in \mathrm{Gal}(L/K) \colon w(\sigma x) > 0 \text{ if } w(x) > 0\}$$

is the *decomposition group* of w over v.

Note that $Z_{w/v}$ is the group of automorphisms of L over K that are continuous in the topology induced by w.

Lemma 5.3.4 *The group $Z_{w/v}$ is isomorphic to $\mathrm{Gal}(L_w/K_v)$.*

Proof Since each $\sigma \in Z_{w/v}$ is continuous, it extends to the completion to give an element of $\mathrm{Gal}(L_w/K_v)$. This gives an embedding of $Z_{w/v}$ into $\mathrm{Gal}(L_w/K_v)$. Conversely, let $\sigma \in \mathrm{Gal}(L_w/K_v)$. Since the topology on L_w is unique, σ and σ^{-1} are both continuous. By Lemma 1.2.8, it follows that $\sigma(\pi_w \mathfrak{o}_w) = \pi_w \mathfrak{o}_w$. Thus the restriction of σ to L lies in $Z_{w/v}$.   □

From this lemma and Lemma 1.2.12, we deduce that the group $Z_{w/v}$ has order $e(w/v)f(w/v)$. In particular, each factor m_w of m over K_v has this degree.

We have seen in the above proof that $\sigma(\pi_w \mathfrak{o}_w) = \pi_w \mathfrak{o}_w$ for $\sigma \in Z_{w/v}$. Therefore, an automorphism in $Z_{w/v}$ induces an automorphism of the residue class field $L(w)$ over $K(v)$.

Lemma 5.3.5 *The map $Z_{w/v} \to \mathrm{Gal}(L(w)/K(v))$ is a surjective homomorphism.*

Proof   Let $x \in \mathfrak{o}_w$ be an element whose class generates $L(w)$ over $K(v)$ and let ϕ be an automorphism of $L(w)$. The polynomial

$$f(X) = \prod_{\sigma \in Z_{w/v}} (X - \sigma x)$$

is defined over $\mathfrak{o}_v$. Let $\bar{f}(X)$ be the polynomial f modulo π_v and let $\bar{x}$ denote the class of x in $L(w)$. Thus $\bar{f}(X) = \prod_\sigma (X - \overline{\sigma x})$. Clearly, $\bar{f}(\bar{x}) = 0$, and hence $\bar{f}(\phi(\bar{x})) = 0$. It follows that $\phi(\bar{x})$ is represented by a root of f. Hence there exists an automorphism in $Z_{w/v}$ that induces ϕ. $\square$

The kernel of the homomorphism of Lemma 5.3.5 is called the *inertia group*

$$T_{w/v} = \{\sigma \in Z_{w/v} : \sigma\xi = \xi \bmod \pi_w \text{ for every } \xi \in \mathfrak{o}_w\}.$$

This group has $e(w/v)$ elements, since $L(w)$ has degree $f(w/v)$ over $K(v)$. The Galois group of $L(w)$ over $K(v)$ is generated by the Frobenius automorphism $x \mapsto x^{q_v}$, where $q_v = q^{\deg v}$ is the cardinality of $K(v)$. Since the homomorphism of Lemma 5.3.5 is surjective, we can find $e(w/v)$ different automorphisms in $\mathrm{Gal}(L/K)$ that induce the Frobenius automorphism of $L(w)$ over $K(v)$. If $T_{w/v}$ is trivial, then this Frobenius automorphism is uniquely determined inside $Z_{w/v}$.

From the preceding discussion, we obtain the next lemma:

Lemma 5.3.6   *Let w and w' be two valuations of L and let σ be an isomorphism $L_{w'} \to L_w$. Then*

$$\sigma(\pi_{w'}\mathfrak{o}_{w'}) = \pi_w\mathfrak{o}_w, \quad Z_{w'/v} = \sigma^{-1}Z_{w/v}\sigma, \quad \text{and} \quad T_{w'/v} = \sigma^{-1}T_{w/v}\sigma.$$

5.4 The Riemann hypothesis for $\mathcal{C}$

Every valuation of K corresponds to $\deg v$ points on $\mathcal{C}$ defined over the finite field $K(v)$. Let $N_\mathcal{C}(n)$ be the number of points on $\mathcal{C}$ with values in $\mathbb{F}_{q^n}$. Since $K(v)$ is a subfield of $\mathbb{F}_{q^n}$ if and only if $\deg v \mid n$, we find that[4] $N_\mathcal{C}(n) = \sum_{\deg v \mid n} \deg v = \sum_{d \mid n} dV_d$. By Example 3.4.3, we obtain

$$N_\mathcal{C}(n) = q^n - \omega_1^n - \cdots - \omega_{2g}^n + 1.$$

Exercise 5.4.1   Describe a procedure to compute $\zeta_\mathcal{C}$ by counting $N_\mathcal{C}(n)$ for only finitely many values of n. What is the minimal number of values of $N_\mathcal{C}$ that you need?

---

[4] See Example 3.4.3 for the definition of V_d.

Exercise 5.4.2   Find a procedure to determine the genus of $\mathcal{C}$ by using finitely many values of $N_\mathcal{C}(n)$.

Recall that the Riemann hypothesis for $\mathcal{C}$ can be formulated as the bound $|\omega_\nu| \leq \sqrt{q}$ for $\nu = 1, \ldots, 2g$. It follows from the Riemann hypothesis that

$$|N_\mathcal{C}(n) - q^n - 1| \leq 2gq^{n/2}.$$

Conversely, we have the following lemma, which states in particular that it suffices to prove this inequality for all even n.

Lemma 5.4.3   *If for every $\varepsilon > 0$ there exists a natural number m such that the inequality*

$$|N_\mathcal{C}(nm) - q^{nm} - 1| \leq Cq^{nm(1/2+\varepsilon)} \tag{5.7}$$

is satisfied for every n, then the Riemann hypothesis holds for $\zeta_\mathcal{C}$.

Proof   Let $\varepsilon > 0$ and suppose that (5.7) is satisfied. By Diophantine approximation, we can find infinitely many n such that $\operatorname{Re}\omega_\nu^{nm} \geq \frac{1}{2}|\omega_\nu|^{nm}$ for $\nu = 1, \ldots, 2g$. Then

$$|N_\mathcal{C}(nm) - q^{nm} - 1| \geq \frac{1}{2}\max_\nu |\omega_\nu|^{nm}.$$

Letting $n \to \infty$, we find that $|\omega_\nu| \leq q^{1/2+\varepsilon}$ for every ν. Since this holds for every $\varepsilon > 0$, we obtain $|\omega_\nu| \leq q^{1/2}$. $\square$

Exercise 5.4.4   Another proof is obtained by considering the radius of convergence of $L'_\mathcal{C}/L_\mathcal{C}$.

5.4.1 The Frobenius flow

The Frobenius automorphism of $\mathbb{F}_q^a$ sends the point x of $\mathcal{C}(\mathbb{F}_q^a)$ to $\phi_q(x)$. We call this the *Frobenius flow* on $\mathcal{C}$. Figure 5.3 depicts the graph

$$Y = \phi_q(X)$$

of this flow in $\mathcal{C} \times \mathcal{C}$.

The intersection with the diagonal

$$\Delta \colon Y = X$$

gives the points (x, x) with $\phi_q(x) = x$. These are the points on $\mathcal{C}$ defined over $\mathbb{F}_q$, and their number is $N_\mathcal{C}(1)$. We assume that there is at least one such point, which we denote by (∞, ∞). We write v_∞ for the corresponding valuation of degree 1 of K.

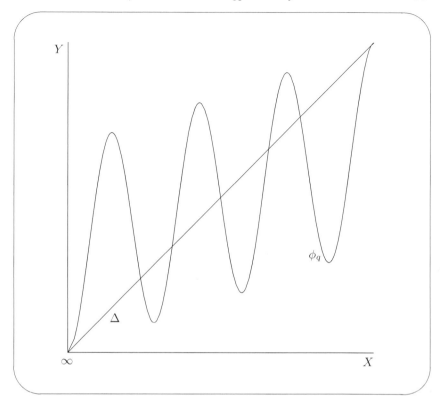

Figure 5.3 The graph of Frobenius intersected with the diagonal.

Remark 5.4.5 The Frobenius automorphism is smooth, since it is a polynomial map. Also, its derivative vanishes, so our intuition says that this map should be constant, or at least locally constant. Being a polynomial of degree q, it seems to be a q-to-one map, but in fact, it is one-to-one. Figure 5.3 emphasizes the smoothness and ignores the injectivity of the Frobenius flow.

The functions defined over $\mathbb{F}_q$ with a pole of order at most m at ∞ and no other poles form an $\mathbb{F}_q$-vector space $L_m = L(m(\infty))$. By Theorem 3.1.9, the dimension of L_m is

$$l_m = l(m(\infty)) = m + 1 - g + l(\mathcal{K} - m(\infty)).$$

We find

$$m + 1 - g \le l_m \le m + 1, \tag{5.8}$$

and

$$l_m = m + 1 - g$$

for $m > \deg \mathcal{K} = 2g - 2$.

Clearly, L_{m+1} contains L_m as a subspace. Also, $l_{m+1} \leq l_m + 1$, since for two functions $f, g \in L_{m+1}$ for which $f \notin L_m$, we can find a constant $\lambda \in \mathbb{F}_q$ such that $g - \lambda f \in L_m$.

Exercise 5.4.6   Given $f \in L_{m+1} \backslash L_m$ and $g \in L_{m+1}$, determine $\lambda \in \mathbb{F}_q$ such that $g \in \lambda f + L_m$.

Let

$$s_1, \ldots, s_{l_m}$$

be a basis for L_m such that $v_\infty(s_{i+1}) < v_\infty(s_i)$, i.e., the order of the pole of s_i at ∞ increases with i. Given $k \geq 0$, we choose coefficients $a_i \in L_k$ to form

$$f(X, Y) = \sum_{i=1}^{l_m} a_i(X) s_i(Y). \qquad (5.9)$$

Remark 5.4.7   For $\mathcal{C} = \mathbb{P}^1$,

$$s_1(Y) = 1, \ s_2(Y) = Y, \ s_3(Y) = Y^2, \ldots,$$

and $f(X, Y)$ is a polynomial in Y with coefficients in X, which we assume to be of degree at most k in X.

This is analogous to a polynomial in Y with integer coefficients that are bounded in size by e^k. Therefore, we call Y the "geometric" coordinate, and X the "arithmetic" coordinate. See Section 5.5 for some speculations with this idea.

We thus obtain a space of functions $f(X, Y)$ on $\mathcal{C} \times \mathcal{C}$ defined over $\mathbb{F}_q$. The restriction of $f(X, Y)$ to the graph of Frobenius is

$$f_{|\phi}(X) = f(X, \phi_q(X)).$$

The map $f \mapsto f_{|\phi}$ is $\mathbb{F}_q$-linear. Note that $s_i(\phi_q(X)) = s_i^q(X)$, since s_i is defined over $\mathbb{F}_q$. Hence $f_{|\phi} \in L_{k+qm}$. That is, $f_{|\phi}$ only has a pole at ∞, of order at most $k + qm$.

Lemma 5.4.8   *For $k < q$, the map $f \mapsto f_{|\phi}$ is injective, and hence an isomorphism onto its image.*

Proof   Let f be given by (5.9), and assume that $f_{|\phi} = 0$. It follows that $\sum_{i=1}^{l_m} a_i s_i^q = 0$ in K. Consider the order of the pole at ∞. If $a_i \neq 0$, then

$$v_\infty(a_i s_i^q) \le q v_\infty(s_i) \le -q + q v_\infty(s_j)$$

for every $j < i$. Further, for $j < i$,

$$v_\infty(a_j s_j^q) \ge -k + q v_\infty(s_j) > -q + q v_\infty(s_j).$$

Hence the pole of the nonzero term of highest order in $f_{|\phi}$ is not cancelled by the pole of any of the other terms. It follows that the highest nonzero term vanishes. This contradiction shows that there is no highest nonzero term, and we conclude that $f = 0$. $\square$

We take the coefficients a_i to be p^μ-th powers, where $p^\mu \mid q$ will be specified later. Thus the coefficients are of the form

$$a_i = b_i^{p^\mu}.$$

Then $f_{|\phi}$ is a p^μ-th power as well.

We choose $b_i \in L_n$, so that $a_i \in L_{p^\mu n}$ and $k = p^\mu n$ in (5.9) and the sentence above (5.9). The space of functions $f(X, Y)$ constructed in this way has dimension $l_n l_m$. To be able to apply Lemma 5.4.8, we assume that

$$k = p^\mu n < q, \tag{5.10}$$

so that also the space of functions $f_{|\phi}$, i.e., the image of the map of Lemma 5.4.8, has dimension $l_n l_m$.

Restricting f to the diagonal, $f_{|\Delta}(X) = f(X, X)$, we obtain the two restriction maps, for $p^\mu n < q$,

$$f \longrightarrow f_{|\Delta}$$
$$\|$$
$$f_{|\phi}$$

where the vertical equality is the isomorphism of Lemma 5.4.8. Using the inverse $f_{|\phi} \mapsto f$, we obtain a linear map $f_{|\phi} \mapsto f_{|\Delta}$.

Assume that $f_{|\Delta} = 0$ and $f_{|\phi} \neq 0$. Then we have a function $f_{|\phi}$ on $\mathcal{C}$ (obtained by restricting a function on $\mathcal{C} \times \mathcal{C}$) that vanishes on the diagonal, except at (∞, ∞). The diagonal intersects the graph of Frobenius at $N_{\mathcal{C}}(1)$ points. Since $f_{|\phi}$ is a p^μ-th power, and (∞, ∞) is the only pole, this function has at least

$$p^\mu \big(N_{\mathcal{C}}(1) - 1\big)$$

zeros, counted with multiplicity. On the other hand, the number of zeros is equal to the order of the pole at ∞. By (5.9) and (5.10), the order of this pole is at most

$$p^\mu n + qm.$$

Hence we find $N_{\mathcal{C}}(1) \leq 1 + n + \frac{q}{p^\mu} m$. By (5.10), $1 + n \leq qp^{-\mu}$, hence we obtain

$$N_{\mathcal{C}}(1) \leq \frac{q}{p^\mu}(m + 1). \tag{5.11}$$

In particular, for given μ, the best bound for $N_{\mathcal{C}}(1)$ is obtained when m is as small as possible.

Example 5.4.9   For genus zero, we take

$$f(X, Y) = X^{p^\mu} - Y^{p^\mu}.$$

Then $f_{|\phi}(X) = X^{p^\mu} - X^{qp^\mu}$ and $f_{|\Delta}(X) = 0$. The function $f_{|\phi}$ has a pole at infinity of order qp^μ, and at least $p^\mu(N_{\mathcal{C}}(1) - 1)$ zeros, counted with multiplicity. Thus the number of points on the projective line over $\mathbb{F}_q$ satisfies $N_{\mathbb{P}^1}(1) \leq q + 1$. In fact, equality holds.

For higher genus, it is much harder to explicitly find a function f such that $f_{|\Delta} = 0$ and $f_{|\phi} \neq 0$. But we can prove the existence of such a function.

Assume that

$$n, m \geq g.$$

Then $l_n l_m \geq (n + 1 - g)(m + 1 - g)$ by (5.8).

The functions $f_{|\Delta}$ lie in $L_{p^\mu n + m}$. To assure the existence of a nontrivial function f such that $f_{|\Delta} = 0$, we choose n and m so that

$$(n + 1 - g)(m + 1 - g) > l_{p^\mu n + m},$$

since then the kernel of $f_{|\phi} \mapsto f_{|\Delta}$ is nontrivial. Since $p^\mu n + m > 2g - 2$, this means that we want

$$(n + 1 - g)(m + 1 - g) > p^\mu n + m + 1 - g,$$

or equivalently,

$$(m + 1 - g - p^\mu)(n - g) > p^\mu g. \tag{5.12}$$

Since we want to choose m as small as possible, we will choose n as large as possible. The largest value for n so that (5.10) is satisfied is

$$n = qp^{-\mu} - 1.$$

We thus obtain from (5.12) a lower bound for m, which we may write as

$$\frac{q}{p^\mu}(m+1) > q + g\left(\frac{q}{p^\mu} + \frac{p^\mu}{1 - (g+1)p^\mu/q}\right). \qquad (5.13)$$

We will choose q and p^μ such that $q > (g+1)p^\mu$. Then we obtain the preliminary estimate

$$\frac{q}{p^\mu}(m+1) > q + g\left(\frac{q}{p^\mu} + p^\mu\right) \geq q + 2g\sqrt{q},$$

with equality only for $q/p^\mu = p^\mu$.

Thus the upper bound for $N_C(1)$ that we can derive from (5.11) is best possible if $p^\mu = \sqrt{q}$. Therefore, we assume that q is an even power of p, as we may by Lemma 5.4.3, and we choose μ such that $p^\mu = \sqrt{q}$.

By (5.13), we find the following condition for m:

$$m + 1 > \sqrt{q} + 2g + \frac{g(g+1)}{\sqrt{q} - (g+1)}.$$

For $q > (g+1)^4$, this inequality is satisfied for

$$m + 1 = \sqrt{q} + 2g + 1.$$

With this choice for m in (5.11), we obtain the following estimate:

Theorem 5.4.10 *For $q > (g+1)^4$, a square, we have*

$$N_C(1) = |C(\mathbb{F}_q)| \leq q + (2g+1)\sqrt{q},$$

where g is the genus of C.

Recall that the above argument depends on the existence of a point ∞ on $C(\mathbb{F}_q)$. If such a point does not exist, then $N_C(1) = 0$ and the inequality for $N_C(1)$ is trivially satisfied.

5.4.2 Frobenius as symmetries

We also need a lower bound for $N_C(1)$. Let

$$C' \longrightarrow C \longrightarrow \mathbb{P}^1 \qquad (5.14)$$

be the Galois cover corresponding to the Galois closure of the function field of C over $\mathbf{q}$. Let G be the Galois group (fundamental group) of the cover $C' \to \mathbb{P}^1$. For $\sigma \in G$, we define

$$N_{C'}(1, \sigma) = \left|\left\{x \in C'(\mathbb{F}_q^a) : x \text{ projects to } \mathbb{P}^1(\mathbb{F}_q) \text{ and } \phi_q(x) = \sigma(x)\right\}\right|.$$

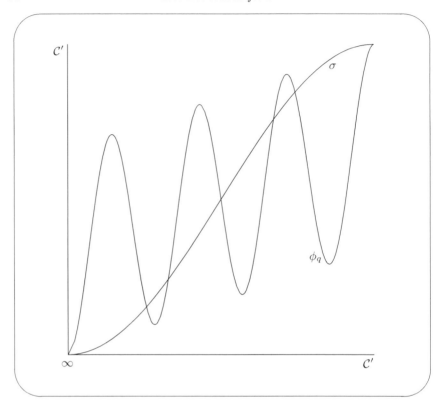

Figure 5.4 The graph of Frobenius intersected with the graph of σ.

Theorem 5.4.11 *For $q > (g+1)^4$, a square, we have*

$$N_{\mathcal{C}'}(1, \sigma) \leq q + (2g' + 1)\sqrt{q},$$

where g' is the genus of $\mathcal{C}'$.

Proof Let X and Y again denote the "arithmetic" and "geometric" coordinates on $\mathcal{C}' \times \mathcal{C}'$; see Figure 5.4. As in Section 5.4.1, we have the restrictions

$$
\begin{array}{ccc}
f & \longrightarrow & f_{|\sigma} \\
\| & & \\
f_{|\phi} & &
\end{array}
$$

where $f_{|\sigma}(X) = f(X, \sigma(X))$ is the restriction of $f(X, Y)$ to the graph of σ and, as before, $f_{|\phi}(X)$ is the restriction to the graph of the Frobenius flow. Clearly, if $f_{|\sigma}$ vanishes, then $f_{|\phi}$ vanishes at the points counted in $N_{C'}(1, \sigma)$. The rest of the argument is as before, applied to C' and the homomorphism $f_{|\phi} \mapsto f_{|\sigma}$.   □

Exercise 5.4.12 Complete the proof of Theorem 5.4.11.

Theorem 5.4.13 *The curve C satisfies the Riemann hypothesis:*

$$\text{if } \zeta_C(\rho) = 0 \text{ then } \operatorname{Re} \rho = 1/2.$$

Proof As in (5.14), let C' be the Galois closure of the cover $C \to \mathbb{P}^1$, with Galois group G. Consider the sum

$$\sum_{\sigma \in G} N_{C'}(1, \sigma).$$

Above every point t of $\mathbb{P}^1(\mathbb{F}_q)$, we have $|G|/e$ points of $C'(\mathbb{F}_q^a)$, where e is the order of ramification of any of the associated valuations in C'. Further, for a point t' of C' above t, there are e different automorphisms in G that induce Frobenius on the residue class field. Hence in the sum, t is counted $|G|$ times. Since $\mathbb{P}^1(\mathbb{F}_q)$ has $q + 1$ points, we obtain

$$\sum_{\sigma \in G} N_{C'}(1, \sigma) = |G|(q + 1).$$

Remark 5.4.14 We have used here the trivially verified fact that the Riemann hypothesis holds for $\mathbb{P}^1$! See also Example 5.4.9.

By Theorem 5.4.11, applied to each summand $N_{C'}(1, \sigma)$ for $\sigma \neq \tau$, we obtain for each $\tau \in G$,

$$N_{C'}(1, \tau) = |G|(q + 1) - \sum_{\sigma \neq \tau} N_{C'}(1, \sigma) \geq q - (|G| - 1)(2g' + 1)\sqrt{q} + |G|.$$

Let H be the subgroup of G of covering transformations that act trivially on C. By the same reasoning as above for $\mathbb{P}^1$, we obtain

$$\sum_{\sigma \in H} N_{C'}(1, \sigma) = |H| N_C(1).$$

It follows that

$$N_C(1) \geq q - (|G| - 1)(2g' + 1)\sqrt{q} + |G|.$$

Combined with the upper bound of Theorem 5.4.10 and Lemma 5.4.3, we deduce the Riemann hypothesis for C.   □

5.5 Comparison with the Riemann hypothesis

Recall that one reason to study Bombieri's proof is to see to what extent Riemann–Roch suffices to prove the Riemann hypothesis, since in the next chapter we will explore a formalism that is essentially equivalent to proving Riemann–Roch. At the moment, it is also fun to explore directly if we could get an idea how to prove the original Riemann hypothesis.

The field of functions on a curve is analogous to the field $\mathbb{Q}$ of rational numbers. Indeed, by Tate's thesis, the Riemann zeta function, and all Dedekind zeta functions, are defined in a manner entirely analogous to (3.12). However, since $\mathbb{Q}$ has no field of constants, there is no analogue of the counting function of points on $\mathcal{C}$. Indeed, for every prime number, we can take the class of f modulo p to obtain a function

$$\mathbb{Q} \longrightarrow \bar{\mathbb{F}}_p = \mathbb{Z}_p/p\mathbb{Z}_p \cup \{\infty\}, \qquad f \longmapsto f(p) = \begin{cases} f + p\mathbb{Z} & \text{if } v_p(f) \geq 0, \\ \infty & \text{if } v_p(f) < 0, \end{cases}$$

and $\log p$ would be the number of points corresponding to this valuation, i.e., the dimension of $\mathbb{F}_p$ over the ground field.

There are (at least) two archimedean analogues: for $f \in \mathbb{R}$, consider

$$f \longmapsto \{-\infty, -1, 0, 1, \infty\},$$

mapping $\ker |\cdot| = \{\pm 1\}$ to itself, the "maximal ideal" $(-1, 1)$ to 0, and the numbers with a "pole" (i.e., $|f| > 1$) to $\pm\infty$, only remembering the sign. This loses a lot of information, but is closely related to the theory of hyperrings (see [CoCo2] and the references therein, and below in connection with meromorphic functions and Nevanlinna theory).

An alternative is to view all of $\mathbb{R}$ as the field of archimedean residue classes and map a rational number f to itself,

$$f \longmapsto f \in \mathbb{R}.$$

There are no unramified extensions of $\mathbb{Q}$, and we cannot see a splitting of the valuation v_p into $\log p$ distinct points on some curve. Therefore there is no Frobenius flow as in Section 5.4.1 on $\operatorname{spec} \mathbb{Z}$, even though in an extension of $\mathbb{Q}$ there are local Frobenius automorphisms as in Section 5.4.2, associated with every prime number.

Remark 5.5.1 The residues of a function are defined using a quotient of the additive group: $\operatorname{res}_P(x)$ depends on the class of $x \in \mathbf{q}_P/\mathbf{o}_P$, and $\operatorname{res}_\infty(x)$ depends on this class modulo $T^{-2}\mathbf{o}_\infty$, and similarly for extensions of $\mathbf{q}$, leading to the introduction of the different.

In the counterpart for $\mathbb{Q}$, the residue of a rational number depends on

$$x \in \mathbb{R}/\mathbb{Z}, \quad \text{or on} \quad x \in \mathbb{Q}_p/\mathbb{Z}_p.$$

Thus the archimedean additive structure is without mystery: $\mathbb{R}$ should simply be viewed as the tangent space at $1 \in \mathbb{F}_1$. And the analogue of the additive group $\mathfrak{o}_P$ is the line $\mathbb{R}$ with the Gaussian measure $e^{-\pi x^2}\, dx$, giving it unit volume.

In contrast, the residue class fields are defined using a quotient of the ring structure:

$$\mathbf{q}(P) = \mathfrak{o}_P/P\mathfrak{o}_P,$$

and if we define it as $\mathbf{q}_P/P\mathfrak{o}_P$, we obtain $\mathbf{q}(P) \cup \{\infty\}$. The units $\mathbf{q}(P)$ are found as $\mathfrak{o}_P^*/(1 + P\mathfrak{o}_P)$.

Part of the mystery can be solved: $1 + P\mathfrak{o}_P$ corresponds to the interval $(0, \infty)$ of positive numbers, but what should be the measure on this space? How can $\mathbb{Z}_p/p\mathbb{Z}_p$ be viewed as an extension of degree $\log p$ of $\mathbb{F}_1$? See [Bor, CoCo4, Den2, Har1, Har2, Har3, Sm] for many ideas.

Since the numbers $\log|x|$, for $x \in \mathbb{Q}$, are dense on the real line, we cannot separate the "point counting function" in the different degrees. Instead, we have the function $\psi(x) = \sum_{p^k \le x} \log p$. The analogue of this function for $\mathcal{C}$ is

$$\psi_{\mathcal{C}}(x) = \sum_{n \le \log_q x} N_{\mathcal{C}}(n) = \sum_{k \deg v \le \log_q x} \deg v,$$

which counts the points on $\mathcal{C}$ with multiplicity $[(\log_q x)/n]$.

To obtain an explicit formula for $\psi_{\mathcal{C}}$, we use

$$N_{\mathcal{C}}(n) = q^n - \omega_1^n - \cdots - \omega_{2g}^n + 1$$

to obtain

$$\psi_{\mathcal{C}}(x) = \frac{q^{[\log_q x]} - 1}{1 - q^{-1}} - \sum_{\nu=1}^{2g} \frac{\omega_{\nu}^{[\log_q x]} - 1}{1 - \omega_{\nu}^{-1}} + [\log_q x].$$

Using that $q^{c[\log_q x]} = x^c q^{-c\{\log_q x\}}$ and the Fourier series

$$q^{-c\{x\}} = (1 - q^{-c}) \sum_{n \in \mathbb{Z}} \frac{e^{2\pi i n x}}{c \log q + 2\pi i n} \quad \text{and} \quad \{x\} = \frac{1}{2} - \sum_{n \ne 0} \frac{e^{2\pi i n x}}{2\pi i n},$$

we obtain a formula for $\psi_{\mathcal{C}}(x)$ as a sum over the zeros and poles of $\zeta_{\mathcal{C}}(s)$,

$$
\psi_{\mathcal{C}}(x) = \frac{1}{\log q}\left(\sum_{n=-\infty}^{\infty}\frac{x^{1+2\pi in/\log q}}{1+2\pi in/\log q} - \sum_{\nu=1}^{2g}\sum_{n=-\infty}^{\infty}\frac{x^{\rho_\nu+2\pi in/\log q}}{\rho_\nu+2\pi in/\log q}\right.
$$
$$
\left.+\sum_{n\neq0}\frac{x^{2\pi in/\log q}}{2\pi in/\log q}+\log x\right)-\frac{1}{2}-\frac{1}{1-q^{-1}}+\sum_{\nu=1}^{2g}\frac{1}{1-\omega_\nu^{-1}}.
$$

Remark 5.5.2 This formula should be compared with the explicit formula of Ingham [In, Theorem 29, p. 77],

$$
\psi(x) = x - \sum_{\rho}\frac{x^\rho}{\rho} - \frac{\zeta'}{\zeta}(0) + \sum_{n=1}^{\infty}\frac{x^{-2n}}{2n},
$$

which expresses $\psi(x)$ as a sum over the zeros and poles of the Riemann zeta function. Indeed, $1+2\pi in/\log q$ and $2\pi in/\log q$ runs over all poles of $\zeta_{\mathcal{C}}$ (for $n\in\mathbb{Z}$), and $\rho_\nu+2\pi in/\log q$ (for $n\in\mathbb{Z}$) runs over all zeros of this function.

Table 5.1 compares the Riemann zeta function with $\zeta_{\mathcal{C}}$. The factor

$$
\zeta_{\mathbb{R}}(s) = \pi^{-s/2}\Gamma(s/2)
$$

completes the Riemann zeta function with the archimedean local zeta function. Maybe $\pi^{-s/2}$ is analogous to q^{gs} and the analogue of the factor at infinity $\frac{q^{-s}}{1-q^{-s}}$ is $\Gamma(s/2)$?

As is pointed out in the introduction, even though the Frobenius flow for the rational numbers is not known, the dual algebraic picture obtained from class field theory is complete. The Frobenius flow on $\mathcal{C}$ corresponds to the action of $q\in\mathbb{A}^*/K^*$ on the space $\mathbb{A}/K^*$ and the Frobenius flow on $\operatorname{spec}\mathbb{Z}$ corresponds to the action of $\mathbb{R}^{>0}\subset\mathbb{A}^*/\mathbb{Q}^*$ on the space $\mathbb{A}/\mathbb{Q}^*$. This will be explored in the next chapter.

To obtain the Riemann hypothesis, we only need the analogue for $\operatorname{spec}\mathbb{Z}$ of the first inequality

$$
N_{\mathcal{C}}(1) \le q + O(\sqrt{q})
$$

of Theorem 5.4.10. As a consequence of the density in the real line of the values $\log|x|$ ($x\in\mathbb{Q}$), it becomes unnecessary to establish the lower bound of Section 5.4.2. That is, we only need to translate the part of Bombieri's proof that depends on the Frobenius flow on $\mathcal{C}$, not the part that depends on the Frobenius symmetries of covers of $\mathcal{C}$.

Table 5.1 Comparison of spec $\mathbb{Z}$ and $\mathcal{C}$.

Rational numbers	Function fields
$\zeta_{\mathbb{Z}}(s) = \zeta_{\mathbb{R}}(s) \sum_{n>0} |n|^{-s}$	$\zeta_{\mathcal{C}}(s) = q^{s(g-1)} \sum_{D \geq 0} |D|^{-s}$
$= \zeta_{\mathbb{R}}(s) \prod_p \frac{1}{1-p^{-s}}$	$= q^{s(g-1)} \prod_v \frac{1}{1-q_v^{-s}}$
Simple poles	
at 1, residue 1	at $1 + k\frac{2\pi i}{\log q}$, res. $\frac{h}{(q-1)\log q}$
at 0, residue -1	at $k\frac{2\pi i}{\log q}$, res. $-\frac{h}{(q-1)\log q}$
Zeros	
at $\rho_n = \frac{1}{2} + i\gamma_n$, $n \in \mathbb{Z}$	at $\rho_\nu = \frac{1}{2} + i\gamma_\nu + k\frac{2\pi i}{\log q}$
	$\left(1 \leq \nu \leq 2g, \, q^{\rho_\nu} = \omega_\nu\right)$
Frobenius as symmetries of a cover	
For every extension of $\mathbb{Q}$	For every cover of $\mathcal{C}$
Frobenius flow	
See below	Acting on points of $\mathcal{C}$
Point counting function	
Unknown	$N_{\mathcal{C}}(n) = |\mathcal{C}(\mathbb{F}_{q^n})|$
Prime counting function	
$\psi(x) = \sum_{p^k \leq x} \log p$	$\psi_{\mathcal{C}}(x) = \sum_{k \deg v \leq \log_q x} \deg v$
Riemann hypothesis: γ_n is real	
$\iff \psi(x) = x + O\left(x^{1/2+\varepsilon}\right)$	$\iff \psi_{\mathcal{C}}(x) = \frac{1}{1-q^{-1}} q^{[\log_q x]}$
	$+ O\left(x^{1/2+\varepsilon}\right)$
	$\iff N_{\mathcal{C}}(n) = q^n + O(q^{n/2})$
$\iff \psi(x) \leq x + O\left(x^{1/2+\varepsilon}\right)$	$\implies \psi_{\mathcal{C}}(x) \leq \frac{1}{1-q^{-1}} q^{[\log_q x]}$
	$+ O\left(x^{1/2+\varepsilon}\right)$
	$\implies N_{\mathcal{C}}(n) \leq q^n + O(q^{n/2})$

We could try to construct a polynomial that vanishes at all prime numbers up to a certain bound. The infinitesimal generator of the shift on the real line, which may be the analogue of the Frobenius flow, is the derivative operator. So we may try to construct such a function that vanishes at the prime numbers to a higher order. Then we want to bound the degree of this polynomial.

The "degree" of a rational prime number is $\log p$, so a prime number should correspond to $\log p$ points on some curve. In Nevanlinna Theory (see, for example, [Hay, LanCh]), the counting function of zeros of a meromorphic function f in the disc of radius r is

$$N_f(0, r) = \sum_{|x| < r \,:\, f(x)=0} \operatorname{ord}_x(f) \log \frac{r}{|x|},$$

so one could consider a function such as $f(z) = \prod_{p \le x}(1 - pz)^{[\log_p x]}$ on the unit disc. Then $N_f(0, 1) = \psi(x)$.

Exercise 5.5.3   Look up the definition of $N_f(x, r)$ in a book on Nevanlinna theory ([LanCh], for example), and verify that $N_f(0, 1) = \psi(x)$.

The main problem is that the arithmetic coordinate cannot be compared to the geometric coordinate: $z \in \mathbb{C}$ is the geometric coordinate and the coefficients of f are integers, i.e., functions (or function values) on the arithmetic coordinate $\operatorname{spec}\mathbb{Z}$. Thus there is no diagonal and we have to force the vanishing of f at the primes in an artificial manner. Unfortunately, we do not know how to bound the Nevanlinna height (i.e., the degree) of this function.

It is interesting to pursue this idea a little further. For $r > 0$, let Δ_r be the disc of radius r with boundary Γ_r, positively oriented. Let

$$f(z) = cz^{\operatorname{ord}(f,0)} + \cdots \tag{5.15}$$

be a meromorphic function with leading coefficient c in its Laurent series at 0.

Nevanlinna theory starts with the Poisson–Jensen formula, which we interpret as a sum over all valuations of the field of meromorphic functions on Δ_r,

$$v_0(f) \log r + \sum_{0<|x|<r} v_x(f) \log \frac{r}{|x|} + \int_{\Gamma_r} v_z(f) \frac{dz}{2\pi i z} = -\log|c|, \tag{5.16}$$

where the valuations are $v_z(f) = -\log|f(z)|$ for each z on the boundary of the disc of radius r, and $v_x(f) = \operatorname{ord}(f, x)$ for each x inside the disc.

Note that the sum (5.16) of the valuations is constant, depending on the function but not on r. This should be compared to the fact that

$$\sum_v v(x) \deg v = 0$$

for every nonzero function in the function field K (see Section 3.1.2). The sum over all valuations does not necessarily vanish, but instead, we find the archimedean valuation of c on the right-hand side of (5.16).

If we consider the subfield of functions for which the Laurent series has rational coefficients, then we can write, with c as in (5.15),

$$\sum_{p \ne \infty} v_p(c) \log p + v_0(f) \log r + \sum_{0<|x|<r} v_x(f) \log \frac{r}{|x|} + \int_{\Gamma_r} v_z(f) \frac{dz}{2\pi i z} = 0,$$

where the first sum is over all p-adic valuations of $\mathbb{Q}$.

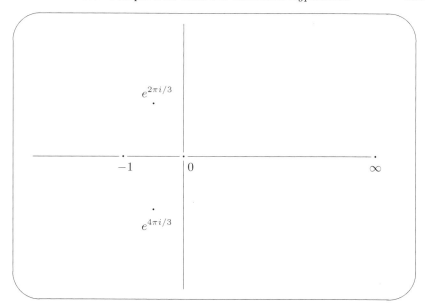

Figure 5.5 Nevanlinna defects.

It is disturbing that the archimedean valuation v_∞ should be excluded from this sum. Following Remark 5.4.7, we interpret this as meaning that the arithmetic and geometric coordinates meet at v_0 (geometrically, $v_0(f)$ is the order of vanishing at zero) and at v_∞ (arithmetically, $v_\infty(f)$ is the size of $f(0)$, or of the first nonvanishing coefficient). Since this is not a double point, we only see v_0.

Observe that $v_0 \log r$ is only a valuation for $r \geq 1$. It is the trivial valuation for $r = 1$. Also, the archimedean valuations have all been pushed to the boundary of the disc, and they are the only nondiscrete valuations.

Remark 5.5.4 For large r, the archimedean valuations are almost nonarchimedean, since for two meromorphic functions f and g, for most points z on the circle Γ_r, $|f(z)+g(z)|$ will be close to $\max\{|f(z)|, |g(z)|\}$, and only on small portions of Γ_r will $f(z)$ and $g(z)$ be comparable in size. More specifically, if the only defects of f/g (in the sense of Nevanlinna theory) are among ∞, 0, -1, $e^{2\pi i/3}$ and $e^{4\pi i/3}$ (see Figure 5.5), then the archimedean valuations behave like nonarchimedean valuations for large r. Moreover, they are almost discrete since $\log|f(z)|$ is close to $[\log|f(z)|]$ if $\log|f(z)|$ is large in absolute value. To formalize this, we could work with the set $\{0, \infty\}$, where ∞ stands for any large number, and 0 stands for a small number, with the following rule for addition:

$$\infty + 0 = \infty, \qquad \infty + \infty = \{0, \infty\}.$$

This leads to the theory of hyperrings of Krasner.

Exercise 5.5.5   The function e^z has two defects, 0 and ∞. Show that, for most z, $|e^z + 1| \approx \max\{|e^z|, 1\}$.

Example 5.5.6   If 2 is a defect of a function, such as $f(z) = 2 + e^z$, then $|f(z) + 1|$ is close to 3 on parts of positive measure on large circles. On those parts, $|f(z) + 1| \neq \max\{|f(z)|, 1\}$, not even approximately.

It seems that by Nevanlinna theory we obtain a connection, albeit a rather loose one, between the geometric valuations v_z and v_x, and the arithmetic valuations v_p, and seemingly, the geometric and arithmetic coordinates intersect at v_0. We consider this the point "at infinity," as in the exposition of the first part of Bombieri's proof in Section 5.4.1.

Trying to copy Bombieri's proof, we could take $s_i = z^{1-i}$ for $i \geq 1$ as the basis of functions that have only a pole at "infinity," and the coefficients are functions on spec $\mathbb{Z}$ with a pole at v_∞ alone, that is, the coefficients are integers b_i. Then

$$f(z) = \sum_{i=1}^{\infty} b_i z^{1-i}.$$

If this series is infinite, then it does not converge for $|z| \leq 1$, hence we assume that f is a rational function. Then the analogue of Lemma 5.4.8 is automatically satisfied, since the coefficients are determined by f. We do not know what the analogue of the choice $b_n = a_n^{p^\mu}$ could be.

6

Shift operators

In this chapter, we compute the trace of the shift on the local additive group K_v^+. This trace gives the Weil term in the explicit formula for the counting function of points on $\mathcal{C}$. Then we give a possible definition and computation of a global trace formula. Even though we suspect that this trace will yield Weil's explicit formula (up to a function that vanishes in the limit), we are not able to complete the computations and consequently, we do not reach our goal of proving Weil positivity.

We encounter two problems: we do not know the correct way to do the cutoff on the adeles, and we cannot always compute the trace when we try to pursue a possible definition.

Computationally, we explore two ways to compute the trace of an operator. The first way is by choosing an orthogonal basis, and then

$$\operatorname{tr} T = \sum \langle e_i, T e_i \rangle.$$

For a suitable choice of the basis, we can compute this sum. The second way is to present T as an integral operator with a kernel. Then the trace of T is the integral over the diagonal of the kernel,

$$\operatorname{tr} T = \int k(x, x) \, dx.$$

If the space where the operator is presented is discrete, then the kernel is simply the matrix k_{ij} of T, and the integral over the diagonal is the usual formula

$$\operatorname{tr} T = \sum_i k_{ii}.$$

This is similar to the first approach, and indeed, if (e_i) is an orthogonal basis, then the matrix coefficients are obtained as $k_{ij} = \langle e_i, T e_j \rangle$. The advantage of the second approach is that no basis needs to be chosen, and in particular, orthogonality does not play a role.

We first compute the shift on the space of coarse idele classes, using both approaches, first with a choice of basis and then using a kernel.

In Section 6.3, we study the local space $K_v^*/\mathfrak{o}_v^*$ (which corresponds to the coarse idele class group and to the group of divisors), and the additive "noncommutative" space $K_v/\mathfrak{o}_v^*$. The cutoff of a function on the additive space is in two steps. First, there is an infrared cutoff, which truncates the support of a function by a multiplication operator, and then an ultraviolet cutoff, which smoothens out the small oscillations of a function by a convolution. The trace gives the local Weil distribution, and this is the strongest indication that the global analogue may give the explicit formula.

The condition that the multiplication and convolution operators commute is as easy to achieve on the global space of adeles as on the local space, and depends on the uncertainty principle for the Fourier transform. The restriction operator E to the ideles connects this with the cutoff on the idele class space. If the additive cutoff would be a projection and the global trace would yield the explicit formula, then we would conclude Weil positivity and hence the Riemann hypothesis for $\mathcal{C}$.

6.1 The Hilbert spaces $\mathcal{Z}$ and $\mathcal{H}$

Recall from Section 3.2 the notation $\ker |\cdot|$ for the group of ideles of vanishing degree. Let $\mathcal{Z}$ be the Hilbert space of square-summable functions on the group of coarse idele classes

$$Z = \mathbb{A}^* / \ker |\cdot|.$$

Clearly, $Z \cong \mathbb{Z}$ by the isomorphism $x \mapsto \deg x$. Thus $\mathcal{Z}$ is the space of functions a on Z such that the norm

$$\|a\|_*^2 = \sum_{n \in Z} |a(n)|^2$$

is finite. The scalar product in this space is

$$\langle a, b \rangle_* = \sum_{n \in Z} \bar{a}(n)\, b(n).$$

We define the *conjugate* of a by $a^*(n) = \bar{a}(1/n)$ (see also Section 4.1).

The *convolution product* is

$$(a * b)(n) = \sum_{k \in Z} a(n/k)b(k).$$

Thus $(a^* * a)(1) = \|a\|_*^2$, and more generally, $(a^* * b)(1) = \langle a, b \rangle_*$.
 The *inversion* is

$$\mathcal{I}_* a(n) = a(1/n).$$

Clearly, $\mathcal{I}_*$ is self-adjoint and unitary. Also, $\mathcal{I}_* U(x) = U(1/x)\mathcal{I}_*$, where
the *unitary shift* over the idele x is defined by the action of Z on itself,

$$U(x)a(n) = a(n/x).$$

It depends only on the norm of x. Its adjoint is $U(1/x)$. For a function h
on Z, the combined shift is $U_h = \sum_x h(x)U(x)$, given by

$$U_h a(n) = \sum_{k \in Z} h(k)a(n/k) = (h * a)(n).$$

Exercise 6.1.1   Verify that $U_h U_k = U_{h*k}$, and that U_{h^*} is the adjoint
of U_h.

Exercise 6.1.2   Let $a \in \mathcal{Z}$ be a nonzero sequence. Then the space
spanned by $U(n)a$ $(n \in Z)$ is a dense subspace of $\mathcal{Z}$.

We also define the Hilbert space $\mathcal{H}$ of functions on the vertical line $i\mathbb{R}$
with period $2\pi i / \log q$, of finite norm

$$\|f\|_\mathcal{H}^2 = \frac{\log q}{2\pi i} \int_0^{2\pi i / \log q} |f(s)|^2 \, ds.$$

For brevity we write the integral as $\|f\|_\mathcal{H}^2 = \int |f(s)|^2 \, ds$. The scalar
product in this space is

$$\langle f, g \rangle_\mathcal{H} = \int \bar{f}(s)g(s) \, ds.$$

The *conjugate* of f is defined as

$$f^*(s) = \bar{f}(-\bar{s}).$$

Using $s = -\bar{s}$ when s is purely imaginary we can write the scalar product
as

$$\langle f, g \rangle_\mathcal{H} = \int f^*(s)g(s) \, ds = \frac{\log q}{2\pi i} \int_\sigma^{\sigma + 2\pi i / \log q} f^*(s)g(s) \, ds,$$

independent of σ, as long as f and g are holomorphic in $\{s \colon |\operatorname{Re} s| \le |\sigma|\}$.

The *unitary shift* in this space, defined by

$$U_{\mathcal{H}}(n)f(s) = |n|^s f(s),$$

only depends on the norm of the idele n, with adjoint $U_{\mathcal{H}}(1/n)$. It satisfies $\mathcal{I}U_{\mathcal{H}}(n) = U_{\mathcal{H}}(1/n)\mathcal{I}$, where the *inversion* in $\mathcal{H}$ is again self-adjoint and unitary,

$$\mathcal{I}f(s) = f(-s).$$

Recall the Mellin transform (4.1),

$$\mathcal{M}a(s) = \sum_{n \in Z} a(n)|n|^s.$$

The Mellin transform is an isomorphism between $\mathcal{Z}$ and $\mathcal{H}$ that intertwines the structure that we have introduced: $\mathcal{M}U(n) = U_{\mathcal{H}}(n)\mathcal{M}$, it preserves inversion, $\mathcal{M}\mathcal{I}_* = \mathcal{I}\mathcal{M}$, and conjugates, $\mathcal{M}(a^*) = (\mathcal{M}a)^*$. The convolution product corresponds to the ordinary product of functions,

$$\mathcal{M}(a * b)(s) = \mathcal{M}a(s)\mathcal{M}b(s).$$

In particular, the combined shift on $\mathcal{H}$ is simply the multiplication operator by $\mathcal{M}h(s)$,

$$\sum_{n \in Z} h(n)U_{\mathcal{H}}(n)f(s) = \sum_{n \in Z} h(n)|n|^s f(s) = \mathcal{M}h(s)f(s). \tag{6.1}$$

For $n \in Z$, we recover $a(n)$ by the inversion formula (4.2), which we write as

$$a(n) = \fint \mathcal{M}a(s)|n|^{-s}\, ds.$$

6.1.1 The truncated shift on $\mathcal{Z}$

The functions δ_n such that $\delta_n(x) = 1$ if and only if $\deg x = n$ (corresponding to $\mathcal{M}\delta_n(s) = q^{-ns}$) clearly form an orthonormal basis for $\mathcal{Z}$.

Given a positive number ω, let M_ω be the projection on the space of functions supported in $\deg x \geq -\omega$. For $n \geq -\omega$, we have $M_\omega \delta_n = \delta_n$, and for $n < -\omega$, $M_\omega \delta_n$ vanishes. For $\theta \geq 0$, the dual projection is

$$\widehat{M_\theta} = \mathcal{I}_* M_\theta \mathcal{I}_*,$$

the projection onto functions supported in $\deg x \leq \theta$. These two projections commute since M_ω and $\widehat{M_\theta}$ are multiplication operators by the indicator functions of $\{x\colon \deg x \geq -\omega\}$ and $\{x\colon \deg x \leq \theta\}$, respectively. This will be confirmed in Section 6.1.2, when we compute the kernel of the operator $\widehat{M}_\theta M_\omega$.

We compute the trace

$$\operatorname{tr}\!\big(\widehat{M}_\theta M_\omega U(a)\big) = \sum_{n=-\infty}^{\infty} \big\langle \delta_n, \widehat{M}_\theta M_\omega U(a)\delta_n \big\rangle.$$

The inversion is self-adjoint, and $\mathcal{I}_*\delta_n = \delta_{-n}$. Also, M_θ is self-adjoint, hence

$$\big\langle \delta_n, \mathcal{I}_* M_\theta \mathcal{I}_* M_\omega U(a)\delta_n \big\rangle = \big\langle M_\theta \delta_{-n}, \mathcal{I}_* M_\omega U(a)\delta_n \big\rangle.$$

This last expression vanishes for $n > \theta$. Also, $U(a)\delta_n = \delta_{n+m}$, where m is the degree of a. We obtain

$$\big\langle \delta_n, \widehat{M}_\theta M_\omega U(a)\delta_n \big\rangle = \begin{cases} \big\langle M_\omega \delta_n, \delta_{n+m} \big\rangle & \text{for } n \leq \theta, \\ 0 & \text{for } n > \theta. \end{cases}$$

This last expression equals $\big\langle \delta_n, \delta_{n+m} \big\rangle$ for $-\omega \leq n \leq \theta$, and vanishes otherwise. Thus we find the trace

$$\operatorname{tr}\!\big(\widehat{M}_\theta M_\omega U(a)\big) = \sum_{-\omega \leq n \leq \theta} \big\langle \delta_n, \delta_{n+m} \big\rangle = \begin{cases} \theta + \omega + 1 & \text{if } m = 0, \\ 0 & \text{if } m \neq 0. \end{cases}$$

6.1.2 The trace using a kernel

It is instructive to also compute this trace using a kernel (that is, the matrix). Clearly, the inversion is given by the kernel

$$\mathcal{I}_* a(x) = \sum_{y \in Z} I_*(x, y)a(y),$$

where the function $I_*(x, y) = 1$ only for $y = 1/x$, and otherwise it vanishes (note that $y = 1/x$ in Z if and only if $\deg y = -\deg x$).

The cutoff is given by the kernel

$$M_\omega a(x) = \sum_{y \in Z} b_\omega(x, y)a(y),$$

where $b_\omega(x, y) = 1$ only for $y = x$ and $\deg x \geq -\omega$. The dual cutoff has kernel $\hat{b}_\theta(x, y)$, which equals 1 only for $y = x$ and $\deg x \leq \theta$, and vanishes otherwise. The combined cutoff has kernel[1] $\delta_{-\omega \leq \deg x \leq \theta}\delta_{y=x}$.

---

[1] For a predicate P, we use the following extension of the Kronecker delta notation: $\delta_P = 1$ if P is true, and $\delta_P = 0$ if P is false.

The unitary shift is given by

$$U(n)a(x) = \sum_{y \in Z} u_n(x, y)a(y),$$

where $u_n(x, y) = \delta_{y=x/n}$ for an idele class $n \in Z$. For the truncated shift, we obtain the kernel $\delta_{-\omega \leq \deg x \leq \theta} \delta_{ny=x}$. That is,

$$\widehat{M_\theta} M_\omega U(n)a(x) = \delta_{-\omega \leq \deg x \leq \theta} \sum_{y \in Z} \delta_{ny=x}\, a(y).$$

Summing over the diagonal $x = y$, we obtain the trace as before,

$$\mathrm{tr}\big(\widehat{M_\theta} M_\omega U(n)\big) = \sum_{-\omega \leq \deg x \leq \theta} \delta_{n=1} = \begin{cases} 0 & \text{if } n \neq 1, \\ \theta + \omega + 1 & \text{if } n = 1. \end{cases}$$

For the combined shift U_h, we find

$$\mathrm{tr}\big(\widehat{M_\theta} M_\omega U_h\big) = (\theta + \omega + 1)h(1). \tag{6.2}$$

Since there are no off-diagonal terms of the truncated shift, we say that there is no interaction.

6.2 Shift operators

To put the computations in Sections 6.3, 6.6, and 6.7 in perspective, we study here the abstract situation of a shift, truncated by an infrared and an ultraviolet cutoff.

Let X be a normed space with a multiplication, a measure dx, and an action by a normed group G by shifts: $U(a)$, for $a \in G$, is the operator

$$U(a)f(x) = |a|^{-\nu/2} f(x/a),$$

where ν is such that

$$\int f(x)\, dx = |a|^\nu \int f(ax)\, dx. \tag{6.3}$$

We assume that the norms on X and G are compatible, $|ax| = |a| \cdot |x|$ for $a \in G$ and $x \in X$. By (6.3), the measure $dx/|x|^\nu$ is invariant for the action of G: the value

$$w = \int_{|x|=q^n} \frac{dx}{|x|^\nu} \tag{6.4}$$

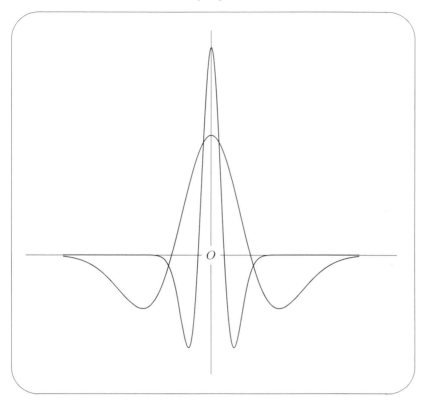

Figure 6.1 The dilated Gaussian.

is independent of n, provided that q^n is the norm of some x. We assume that q is such that exactly the integer powers of q are norms. We write

$$v(x) = -\log_q |x|$$

for the corresponding negative exponent.

Depending on X, there may also exist elements with vanishing norm. In that case, we assume $\nu = 1$, so that the set $\{x \colon |x| = 0\}$ has measure zero.

Example 6.2.1 If $X = G$, we let dx be a Haar measure on X, and take $\nu = 0$. And if X is a field with additive Haar measure dx and G is the multiplicative group of X, then we take $\nu = 1$, in which case we call $U(a)$ a *dilation* operator (which is a unitary operator in $L^2(X)$).

Figure 6.1 depicts the graph of the function $(1 - 2\pi t^2)\, e^{-\pi t^2}$ dilated by the factor $1/3$. See Chapter 7 for some information about this function.

Let $\mathcal{F}$ be the "Fourier transform" on functions on X, given by a kernel,

$$\mathcal{F}f(x) = \int \chi(xy)f(y)\,dy.$$

We assume that χ is real-valued and $\int \chi(xz)\chi(zy)\,dz = \delta_{x=y}$, so that $\mathcal{F}$ is self-adjoint and unitary.

Exercise 6.2.2 Check that $\mathcal{F}U(a) = U(a^{-1})\mathcal{F}$.

Given two positive integers ω and θ, let B_ω be the cutoff on X,

$$B_\omega f(x) = \delta_{|x| \le q^\omega} f(x), \tag{6.5}$$

and let $\widehat{B}_\theta$ be the dual cutoff $\mathcal{F}B_\theta\mathcal{F}$. We also use the notation $B_\omega(x)$ for the function $\delta_{|x| \le q^\omega}$ and $\int_{|x| \le q^\omega} f(x)\,dx$ for $\int B_\omega f(x)\,dx$.

Lemma 6.2.3 *The kernel of $\widehat{B}_\theta B_\omega U(a)$ is given by $|a|^{\nu/2}k(x, ay)$, where*

$$k(x, y) = B_\omega(y)\int_{|z| \le q^\theta} \chi(xz)\chi(zy)\,dz.$$

Proof We compute that $\mathcal{F}B_\omega f(x)$ is given by $\int_{|y| \le q^\omega} \chi(xy)f(y)\,dy$. Hence

$$\mathcal{F}B_\theta\mathcal{F}B_\omega f(x) = \int_{|z| \le q^\theta} \int_{|y| \le q^\omega} \chi(xz)\chi(zy)f(y)\,dy\,dz.$$

Thus the kernel of $\widehat{B}_\theta B_\omega$ is $k(x, y)$. Applying this to $U(a)f$, we find

$$\widehat{B}_\theta B_\omega U(a)f(x) = |a|^{-\nu/2}\int k(x, y)f(y/a)\,dy.$$

Substituting ay for y yields the kernel $|a|^{\nu/2}k(x, ay)$ for this operator. $\square$

Let H be the Hilbert space of functions on X that depend only on $|x|$. In all our examples, the image of the cutoff $\widehat{B}_\theta B_\omega$ is finite dimensional, so the operator $\widehat{B}_\theta B_\omega U(a)$ is of trace class.

Lemma 6.2.4 *Let T be a trace-class operator on $L^2(X)$ given by a kernel, $Tf(x) = \int \kappa(x, y)f(y)\,dy$. Then the trace of T is the integral of the kernel over the diagonal,*

$$\operatorname{tr} T = \int \kappa(x, x)\,dx.$$

Proof Let $e_1, e_2, \ldots$ be an orthonormal basis for $L^2(X)$. Then

$$\operatorname{tr} T = \sum_i \langle e_i, T e_i \rangle.$$

The function $y \mapsto \bar{\kappa}(x, y)$ can be written as $\bar{\kappa}(x, y) = \sum_j \bar{k}_j(x) e_j(y)$ for certain functions $\bar{k}_j(x)$. Then $T e_i(x) = k_i(x)$, so that $\operatorname{tr} T = \sum_i \langle e_i, k_i \rangle$. The integral over the diagonal of the kernel of T is $\int \sum_i k_i(x) \bar{e}_i(x) \, dx$, which equals $\sum_i \langle e_i, k_i \rangle$ as well. This proves the lemma. $\qquad \square$

Recall the value w of formula (6.4).

Lemma 6.2.5 *The trace of $\widehat{B}_\theta B_\omega U(a)$ is*

$$\operatorname{tr}\big(\widehat{B}_\theta B_\omega U(a)\big) = w|a|^{-\nu/2} \int_{v(x) \geq -\theta-\omega} (\theta + \omega + 1 + v(x)) \chi(x/a) \chi(x) \, dx.$$

Proof The trace is found as the integral over the diagonal,

$$\operatorname{tr}\big(\widehat{B}_\theta B_\omega U(a)\big) = |a|^{\nu/2} \int k(x, ax) \, dx = |a|^{-\nu/2} \int k(x/a, x) \, dx.$$

By Lemma 6.2.3, we find $|a|^{-\nu/2} \iint B_\theta(z) B_\omega(x) \chi(xz/a) \chi(zx) \, dz \, dx$ for the trace. Since the set of x with $|x| = 0$ is neglegible, we can substitute z/x for z and change the order of integration to obtain

$$\operatorname{tr}\big(\widehat{B}_\theta B_\omega U(a)\big) = |a|^{-\nu/2} \int \left(\int B_\theta(z/x) B_\omega(x) \frac{dx}{|x|^\nu} \right) \chi(z/a) \chi(z) \, dz.$$

The inner integral equals $(\theta + \omega + 1 + v(z))w$ for $v(z) \geq -\theta - \omega - 1$, and vanishes otherwise. $\qquad \square$

For example, the conditions of the lemmas in this section are satisfied for the space $\mathcal{Z}$, with $X = Z$ is the coarse idele class group, $\chi(x) = \delta_{|x|=1}$, and $\nu = 0$. We find the kernel of the truncated shift as in Section 6.1.2, and the trace is

$$\operatorname{tr}\big(\widehat{M}_\theta M_\omega U(n)\big) = \int (\theta + \omega + 1 + v(x)) \delta_{|x/n|=1} \delta_{|x|=1} \, d^* x,$$

which yields the trace as before.

6.2.1 Averaging spaces

For an operator on a space obtained by an averaging procedure, it is necessary to take the same average of the kernel.

As an example, consider the space of functions on the two-element group $G = \mathbb{Z}/2\mathbb{Z}$ with the operator S with kernel $k(x, y) = \delta_{x+y=1}$. Clearly, the matrix of S is

$$\begin{bmatrix} 0 & 1 \\ 1 & 0 \end{bmatrix}$$

on the basis $\{\delta_0, \delta_1\}$, where $\delta_x(y) = \delta_{x=y}$ for $x, y \in G$. Hence the trace of S vanishes, which we also find using the kernel,

$$\operatorname{tr} S = \sum_{x=0}^{1} k(x, x) = 0.$$

The subspace of G-invariant functions is one-dimensional, spanned by the function $\delta_0 + \delta_1$. It has an induced action of S, denoted by $\bar{S}$, which reduces to the identity operator on this one-dimensional space: the kernel is compatible with the action of G in the sense that

$$k(x + a, y) = k(x, a + y). \tag{6.6}$$

It follows that if f is G-invariant, then so is Sf, given by

$$Sf(x) = \sum_{y=0}^{1} k(x, y) f(y) = f(1 - x),$$

which coincides with $f(x)$ since f is invariant. Hence $\operatorname{tr} \bar{S} = 1$.

We see that the induced action is given by the same kernel. But if we compute the trace of the induced action using this kernel, just as we did above, we find that the trace vanishes. Instead, to retrieve the correct value of the trace, we first need to take the average of the kernel as follows.

Let dx be the measure on G that gives each point measure $1/2$. Then the averaging operator

$$\oint f(x) = \int_G f(x + a)\, da = \frac{1}{2} f(x) + \frac{1}{2} f(1 - x)$$

projects a function onto the space of G-invariant functions. For an invariant function,

$$Sf(x) = S\!\!\oint\!f(x) = \sum_{y=0}^{1} \int_G k(x, y) f(a + y)\, da$$

$$= \sum_{y=0}^{1} \int_G k(x + a, y) f(y)\, da = \frac{1}{2}(f(0) + f(1)).$$

Hence the average kernel of the induced action is $\int_G k(x+a, y)\, da$, which equals $1/2$ for every $x, y \in G$. With this kernel, we obtain the correct value of the trace.

The reason for this discrepancy is that, because of (6.6), the same kernel can be used for functions on G and for S-invariant functions. In the integral over the diagonal, the information of the space on which the operator acts is lost.

This observation will be important when we consider the trace on the space of functions on $\mathbb{A}^*/\mathfrak{o}^*$ or on $\mathbb{A}/\mathfrak{o}^*$. Instead of taking the quotient space, we can instead consider $\mathfrak{o}^*$-invariant functions on $\mathbb{A}^*$ or $\mathbb{A}$, provided that we average the kernel over $\mathfrak{o}^*$.

6.3 Local trace

The Weil distribution is obtained as the trace of the shift, cut off on the additive space. The operator

$$A_v f(x) = |x|_v^{-1/2} f(x): \quad L^2\left(K_v^*/\mathfrak{o}_v^*, d_v^*\right) \longrightarrow L^2\left(K_v/\mathfrak{o}_v^*, d_v\right).$$

lifts a function on $K_v^*/\mathfrak{o}_v^*$ to a function on $K_v/\mathfrak{o}_v^*$. Its inverse is

$$E_v f(a) = |a|_v^{1/2} f(a),$$

which makes it a function on the multiplicative group again. Thus the spaces $L^2(K_v^*/\mathfrak{o}_v^*, d_v^*)$ and $L^2(K_v/\mathfrak{o}_v^*, d_v)$ are isomorphic, and we do not encounter the problems of the semi-local and global situations, see Sections 6.5–6.7.

The *infrared cutoff* is given by multiplication by a function with large support,[2]

$$M_\Omega f(x) = \Omega(x) f(x),$$

where $\Omega = \mathfrak{p}_v^{-\omega}$. The *ultraviolet cutoff* restricts the small-scale behaviour of a function by the convolution with the Fourier transform of θ,

$$C_{\mathcal{F}_v \theta} f(x) = \int_{K_v} \mathcal{F}_v \theta(x - y) f(y)\, d_v y.$$

---

[2] We use the same notation for the function and the divisor defining this function. Similarly, for aesthetic reasons, we use the same notation for the function θ defining the ultraviolet cutoff as for the degree of the associated divisor (not for the divisor itself). See also Remark 5.1.1.

We take $\theta = \mathbf{p}_v^{-\theta}$ for some $\theta \geq 0$, so that $\mathcal{F}_v\theta$ is locally constant at distances in $\pi_v^{\theta-k_v} \mathbf{o}_v$. These two cutoffs commute if

$$\omega + \theta \geq k_v, \tag{6.7}$$

by the following lemma.

Let supp(f) denote the support of a function, and const(f) the set of distances at which f is constant:

$$\varepsilon \in \mathrm{const}(f) \text{ if and only if } f(x+\varepsilon) = f(x) \text{ for all } x \in K_v.$$

Clearly, const$(f) \subseteq$ supp(f).

Lemma 6.3.1 *Let f and g be two functions on $K_v/\mathbf{o}_v^*$. Then M_f and C_g commute if and only if* supp$(g) \subseteq$ const(f).

Proof For a function φ, $M_f C_g \varphi$ is the function

$$M_f C_g \varphi(x) = f(x) \int_{K_v} g(x-y)\varphi(y)\, d_v y,$$

whereas $C_g M_f \varphi$ is given by $C_g M_f \varphi(x) = \int g(x-y)f(y)\varphi(y)\, d_v y$. Hence the commutator of M_f and C_g has kernel $(f(x) - f(y))g(x-y)$.

Suppose that supp$(g) \subseteq$ const(f). We show that the kernel vanishes. Let $\varepsilon \in$ supp(g). Then $\varepsilon \in$ const(f), and hence $f(x+\varepsilon) = f(x)$ for all x. And for $\varepsilon \notin$ supp(g), $g(\varepsilon) = 0$. It follows that $f(x) - f(y)$ or $g(x-y)$ vanishes for all x, y.

On the other hand, if M_f and C_g commute, then the kernel of the commutator vanishes identically. If $\varepsilon \in$ supp(g), then $g(\varepsilon) \neq 0$. Hence $f(x) - f(x+\varepsilon)$ must vanish for all x, so that $\varepsilon \in$ const(f). We conclude that supp$(g) \subseteq$ const(f). $\square$

The following lemma relates the ultraviolet cutoff on the additive space with the abstract theory of Section 6.2.

Lemma 6.3.2 *The Fourier transform intertwines convolution with multiplication,* $C_{\mathcal{F}_v\theta} = \mathcal{F}_v M_\theta \mathcal{F}_v^{-1}$.

Proof Taking $\mathcal{F}_v^{-1}$ on both sides, we compute

$$\left(\mathcal{F}_v^{-1} C_{\mathcal{F}_v\theta} f\right)(x) = \int \chi_v(-xy) C_{\mathcal{F}_v\theta} f(y)\, d_v y$$

$$= \iint \chi_v(-xy) \mathcal{F}_v\theta(y-z) f(z)\, d_v z\, d_v y.$$

We interchange the order of integration and substitute $y + z$ for y to obtain

$$\left(\mathcal{F}_v^{-1} C_{\mathcal{F}_v \theta} f\right)(x) = \int f(z) \chi_v(-xz) \left(\int \chi_v(-xy) \mathcal{F}_v \theta(y) \, d_v y \right) d_v z,$$

which equals $\theta(x) \mathcal{F}_v^{-1} f(x)$. We conclude that $\mathcal{F}_v^{-1} C_{\mathcal{F}_v \theta} = M_\theta \mathcal{F}_v^{-1}$. $\square$

6.3.1 Order of the cutoffs

In the semi-local and global case, it will be necessary to apply these operators in a particular order, to obtain a lift to the additive space (respectively, K_v, $\mathbb{A}_S$ or $\mathbb{A}$) and a cutoff in several steps: a function on $Z = \mathbb{A}^* / \ker |\cdot|$ could be lifted to a function on $\mathbb{A}$ simply by

$$x \longmapsto f(x \ker |\cdot|),$$

but this function is not compactly supported, since an idele of a certain degree can have a component of large negative degree, and it is also not locally constant (on the adeles), since such an idele can have a component of large positive degree. These problems already come up in the semi-local case if S contains at least two valuations, but are not an issue in the local case, as we now explain.

The first step of lifting a function from the space of coarse idele classes is to restrict the support of a lift: M_Ω maps a function on the space of degrees $K_v^* / \ker |\cdot|_v$ to a function on the space of divisors $K_v^* / \mathfrak{o}_v^*$ with the multiplicative measure d_v^* by restricting the choice of divisor of a particular degree. Since in the local case, $\ker |\cdot|_v = \mathfrak{o}_v^*$, there is no choice involved.

Then A_v changes the measure to the additive measure $|x|_v \, d_v^*$ on the space of divisors.

The third step is to "thicken" a function on the group of divisors to a function on the additive space K_v (which is $\mathfrak{o}_v^*$-invariant) by convolving it with a function that will smoothen out its small scale oscillations. In the local case, this thickening only involves a convolution with a test function that is locally constant at a certain distance.

This will be explained in full detail in Section 6.5. For now, we observe that the lift–cutoff combination must be performed in the order $C_{\mathcal{F}_v \theta} A_v M_\Omega$ and that M_Ω commutes with A_v and $C_{\mathcal{F}_v \theta}$ with our choice of Ω and θ.

6.3.2 Direct computation

First, we compute the trace of $E_v C_{\mathcal{F}_v \theta} A_v M_\Omega U(a)$ by choosing an orthogonal basis. Let δ_n be the function on $K_v^* / \mathfrak{o}_v^*$ defined by

$$\delta_n(x) = \delta_{v(x)=n}.$$

This is similar to Section 6.1.1. The difference is that here $v(x) = 1$ corresponds to a divisor of degree $\deg v$ and not of degree 1. Clearly, the functions δ_n are orthogonal and $\|\delta_n\|_* = 1$. We compute

$$E_v C_{\mathcal{F}_v \theta} A_v M_\Omega U(a)\delta_n.$$

The function Ω is supported on $v(x) \geq -\omega$, hence for $m \geq -\omega$ we find $M_\Omega \delta_m = \delta_m$. This function is then made additive by $A_v \delta_m = q_v^{m/2}\delta_m$. Writing $a = \pi_v^l$ (so that $|a|_v = q_v^{-l}$), the shift is found as $U(a)\delta_n = \delta_{n+l}$. We obtain $A_v M_\Omega U(a)\delta_n = A_v M_\Omega \delta_{n+l}$, and, with $m = n + l$,

$$A_v M_\Omega \delta_m = \begin{cases} q_v^{m/2}\delta_m & \text{if } m \geq -\omega, \\ 0 & \text{if } m < -\omega. \end{cases}$$

Further, δ_m is locally constant at distances in $\pi_v^{m+1}\mathfrak{o}_v$, hence $\mathcal{F}_v^{-1}\delta_m$ is supported on $\pi_v^{-k_v-m-1}\mathfrak{o}_v$, by Theorem 2.1.15. Hence

$$M_\theta \mathcal{F}_v^{-1}\delta_m = \mathcal{F}_v^{-1}\delta_m,$$

and it follows that $\mathcal{F}_v M_\theta \mathcal{F}_v^{-1}\delta_m = \delta_m$ for $-k_v - m - 1 \geq -\theta$.

For $-k_v - m \leq -\theta$, we compute

$$\mathcal{F}_v^{-1}\delta_m = q_v^{-k_v/2-m}\left(\mathfrak{p}_v^{-k_v-m} - q_v^{-1}\mathfrak{p}_v^{-k_v-m-1}\right),$$

and $M_\theta \mathcal{F}_v^{-1}\delta_m = q_v^{-k_v/2-m}(1 - q_v^{-1})\mathfrak{p}_v^{-\theta}$. Taking the Fourier transform again, we find that

$$\mathcal{F}_v M_\theta \mathcal{F}_v^{-1}\delta_m = \begin{cases} \delta_m & \text{for } m < \theta - k_v, \\ q_v^{\theta-k_v-m}(1 - q_v^{-1})\mathfrak{p}_v^{\theta-k_v} & \text{for } m \geq \theta - k_v. \end{cases}$$

Finally, $E_v \delta_m = q_v^{-m/2}\delta_m$ and $E_v \mathfrak{p}_v^{\theta-k_v}(x) = \sqrt{|x|_v}\,\mathfrak{p}_v^{\theta-k_v}(x)$.

Putting everything together, we find that δ_m is truncated to

$$E_v C_{\mathcal{F}_v \theta} A_v M_\Omega \delta_m = \begin{cases} 0 & \text{for } m < -\omega, \\ \delta_m & \text{for } -\omega \leq m < \theta - k_v, \\ Q\sqrt{|\cdot|_v}\,\mathfrak{p}_v^{\theta-k_v} & \text{for } m \geq \theta - k_v, \end{cases}$$

where Q is the factor $q_v^{\theta-k_v-m/2}(1 - q_v^{-1})$.

The trace of an operator T is $\sum_n \langle \delta_n, T\delta_n \rangle$. We apply the above formula with $m = n + l$. For $a \neq 1$ (i.e., $l \neq 0$), $\langle \delta_n, \delta_{n+l} \rangle$ vanishes, and we find

$$(1 - q_v^{-1})q_v^{\theta - k_v - l/2} \sum_{n=\theta - k_v - l}^{\infty} q_v^{-n/2} \langle \delta_n, \sqrt{|\cdot|_v} \, \mathfrak{p}_v^{\theta - k_v} \rangle$$

for the trace of $E_v C_{\mathcal{F}_v \theta} A_v M_\Omega U(a)$. The sum starts at $\theta - k_v$ if $l > 0$, and has value $q_v^{-l/2}$. If $l < 0$, the sum starts at $\theta - k_v - l$, with value $q_v^{l/2}$. Hence we find the value

$$\text{tr}(E_v C_{\mathcal{F}_v \theta} A_v M_\Omega U(a)) = \min\{|a|_v^{1/2}, |a|_v^{-1/2}\} \tag{6.8}$$

for the trace.

On the other hand, for $l = 0$, the sum for $m \geq \theta - k_v$ evaluates to 1, and the value of the trace is

$$\text{tr}(E_v C_{\mathcal{F}_v \theta} A_v M_\Omega U(a)) = \theta + \omega - k_v + 1. \tag{6.9}$$

We summarize these two formulas using the combined shift over a test function,

$$U_h = \sum_{n \in K_v^* / \mathfrak{o}_v^*} h(n) U(n). \tag{6.10}$$

Then (6.8) and (6.9) can be combined in the formula

$$\text{tr}(E_v C_{\mathcal{F}_v \theta} A_v M_\Omega U_h) = (\theta + \omega + 1 - k_v)h(1) + W_v(h),$$

where $W_v(h)$ is the Weil distribution as given in (4.5).

In the next section, we will see how taking the function h into account from the beginning allows us to recover a formula that is very close to Weil's original expression for the local terms in the explicit formula. See Remark 4.3.1.

6.3.3 Trace using the kernel

We compute the kernel of the combined shift U_h of (6.10), cut off on the additive space.

We could write $C_{\mathcal{F}_v \theta} = \mathcal{F}_v M_\theta \mathcal{F}_v^{-1}$ (see Lemma 6.3.2), and have a simple application of Lemma 6.2.5, but we choose to present the computation in a different way.

For a function f on $K_v^* / \ker |\cdot|_v$,

$$A_v M_\Omega U_h f(y) = \Omega(y) \sum_n h(n) f(y/n) |y|_v^{-1/2},$$

where the summation is over $K_v^* / \ker |\cdot|_v$. Convolving this with $\mathcal{F}_v \theta$, we obtain by (2.9),

$$C_{\mathcal{F}_v \theta} A_v M_\Omega U_h f(x)$$

$$= \int \mathcal{F}_v \theta(x - y) \Omega(y) \sum_n f(y/n) h(n) |y|_v^{-1/2} \, d_v y$$

$$= \kappa_*^+ \int \Omega(y) \oint \sum_n f(y/(\varepsilon n)) h(n) \mathcal{F}_v \theta(x - y) |y|_v^{1/2} \, d_v^* \varepsilon \, d_v^* y,$$

writing $\oint$ for the average over $\mathfrak{o}_v^*$ (taking this average is necessary, see Section 6.2.1). Substituting $\varepsilon n y$ for y and applying E_v, we find

$$E_v C_{\mathcal{F}_v \theta} A_v M_\Omega U_h f(m)$$

$$= \kappa_*^+ \int f(y) \sum_n \Omega(ny) h(n) \oint \mathcal{F}_v \theta(m - \varepsilon n y) |mny|_v^{1/2} \, d_v^* \varepsilon \, d_v^* y.$$

We have found the kernel of $E_v C_{\mathcal{F}_v \theta} A_v M_\Omega U_h$: for $m, y \in K_v^* / \ker |\cdot|_v$,

$$k(m, y) = \kappa_*^+ \sum_n \Omega(ny) h(n) \oint \mathcal{F}_v \theta(m - \varepsilon n y) |mny|_v^{1/2} \, d_v^* \varepsilon.$$

We change the multiplicative measure $\kappa_*^+ d_v^* \varepsilon$ on the set $|\varepsilon|_v = 1$ to $d_v \varepsilon$ and write the Fourier transform of θ as an integral, to find

$$k(m, y) = \sum_n \Omega(ny) h(n) \int_{\mathfrak{o}_v^*} \int \chi_v((m - \varepsilon n y)x) \theta(x) |mny|_v^{1/2} \, d_v x \, d_v \varepsilon.$$

The integral over ε is the Fourier transform $\mathcal{F}_v(\mathfrak{o}_v^*)(nyx)$. After substituting $x/(ny)$ for x, we obtain our expression for the kernel $k(m, y)$:

$$k(m, y)$$

$$= \sum_n \Omega(ny) h(n) \int \chi_v(mx/(ny)) \mathcal{F}_v(\mathfrak{o}_v^*)(x) \theta(x/(ny)) |ny/m|_v^{-1/2} \, d_v x.$$

The trace is the sum over the diagonal $m = y$,

$$\sum_{m \in K_v^* / \ker |\cdot|_v} k(m, m)$$

$$= \sum_{n,m} \Omega(nm) h(n) |n|_v^{-1/2} \int \chi_v(x/n) \mathcal{F}_v(\mathfrak{o}_v^*)(x) \theta(x/(nm)) \, d_v x.$$

Substituting m/n for m, we obtain

$$\sum_m k(m, m) = \sum_n h(n) |n|_v^{-1/2} \int \sum_m \theta(x/m) \Omega(m) \chi_v(x/n) \mathcal{F}_v(\mathfrak{o}_v^*)(x) \, d_v x.$$

Now $\theta(x/m)\Omega(m)$ vanishes unless $v(x) - v(m) \geq -\theta$ and $v(m) \geq -\omega$. Hence for $v(x) \geq -\theta - \omega - 1$, the sum over m equals $\theta + \omega + 1 + v(x)$, and vanishes otherwise. The integral is restricted by the support of $\mathcal{F}_v(\mathfrak{o}_v^*)$: by (3.2),

$$\mathcal{F}_v(\mathfrak{o}_v^*) = q_v^{-k_v/2}\left(\mathfrak{p}_v^{-k_v} - q_v^{-1}\mathfrak{p}_v^{-k_v-1}\right),$$

hence $v(x) \geq -k_v - 1$. This implies $v(x) \geq -\theta - \omega - 1$, since we need to assume that $\theta + \omega \geq k_v$ (see (6.7)).

We split the value $\theta + \omega + 1 + v(x)$ into $\theta + \omega - k_v$ and $k_v + 1 + v(x)$. This splits the trace into two expressions, of which the first is

$$(\theta + \omega - k_v) \sum_n h(n)|n|_v^{-1/2} \int \chi_v(x/n)\mathcal{F}_v(\mathfrak{o}_v^*)(x)\, d_v x.$$

This integral is a Fourier transform, yielding the value of the indicator function of $\mathfrak{o}_v^*$ at $1/n$, that is, 1 if $n \in \mathfrak{o}_v^*$ and 0 otherwise. Hence we obtain the value $(\theta + \omega - k_v)h(1)$.

In the second expression, $k_v + 1 + v(x)$ vanishes if $v(x) = -k_v - 1$, hence we can restrict the integral to $v(x) \geq -k_v$. On this set, $\mathcal{F}_v(\mathfrak{o}_v^*)$ is constant, equal to $\mu = q_v^{-k_v/2}(1 - q_v^{-1})$. The second expression is therefore

$$\mu \sum_n h(n)|n|_v^{-1/2} \int_{v(x) \geq -k_v} (k_v + 1 + v(x))\chi_v(x/n)\, d_v x.$$

Lemma 6.3.3 *The Fourier transform of $x \mapsto \mu(k_v + 1 + v(x))$ on $v(x) \geq -k_v$ is $\min\{1, 1/|y|_v\}$.*

Proof First note that the function can be written as $f = \mu \sum_{l=0}^{\infty} \mathfrak{p}_v^{l-k_v}$. Using $\mu q_v^{k_v/2} = 1 - q_v^{-1}$, we find

$$\mathcal{F}_v f = (1 - q_v^{-1}) \sum_{l=0}^{\infty} q_v^{-l}\mathfrak{p}_v^{-l}.$$

For $v(y) \geq 0$, $\mathfrak{p}_v^{-l}(y) = 1$ for every $l \geq 0$, hence $\mathcal{F}_v f(y) = 1$. For $v(y) \leq 0$, the sum starts at $-v(y)$, with value $q_v^{v(y)} = 1/|y|_v$. □

By the lemma, with $1/n$ for y, we obtain the value

$$\sum_n \min\{1, |n|_v\}h(n)|n|_v^{-1/2}$$

for the second expression, which equals $h(1) + W_v(h)$. Hence the trace is given by

$$\mathrm{tr}(E_v \mathcal{F}_v M_\theta \mathcal{F}_v^{-1} A_v M_\Omega U_h) = (\theta + \omega + 1 - k_v)h(1) + W_v(h),$$

thus recovering the result of Section 6.3.2.

See also Section 6.7, Lemma 6.7.1, for the following exercise:

Exercise 6.3.4   Compute the trace on the space $\mathcal{H}_v$ of periodic functions with period $2\pi i/\log q_v$. That is, find the kernel of the operator

$$\mathcal{M}_v E_v C_{\mathcal{F}_v \theta} A_v M_\Omega U_h \mathcal{M}_v^{-1},$$

where for $f\colon K_v^*/\ker|\cdot|_v \to \mathbb{C}$, the Mellin transform on $\mathcal{Z}_v$ is defined by

$$\mathcal{M}_v f(s) = \sum_{n \in K_v^*/\ker|\cdot|_v} f(n)|n|_v^s.$$

Remark 6.3.5   In the global explicit formula (4.6) of Section 4.3, each Weil term occurs with multiplicity $\deg v$. This multiplicity comes from the fact that the space of periodic functions in q^s (with period $2\pi i/\log q$) contains $\deg v$ copies of the space of functions in q_v^s. We see a similar phenomenon in Selberg's trace formula, where $\deg v$ corresponds to the length of a geodesic.

6.4 How to prove the Riemann hypothesis for $\mathcal{C}$?

We have seen how the trace of the local shift cut off on the additive space gives the Weil distribution. Let us check that $P = E_v C_{\mathcal{F}_v \theta} A_v M_\Omega$ is an orthogonal projection. From its kernel, we see that P is self-adjoint. Further, since

$$A_v E_v \text{ is the identity,}$$

and

$$M_\Omega \text{ commutes with both } A_v \text{ and } C_{\mathcal{F}_v \theta},$$

it follows that $P^2 = E_v C_{\mathcal{F}_v \theta}^2 M_\Omega^2 A_v$. Since $\Omega^2 = \Omega$ and $\theta^2 = \theta$, we conclude that $P^2 = P$.

Globally, we want to reason as follows: let Ω be a function on the adeles supported in $\{\deg x \geq -\omega\}$, and θ such a function in $\{\deg x \geq -\theta\}$, giving the two cutoffs M_Ω and $C_{\mathcal{F}\theta}$. These cutoffs should commute, hence we require, as in the local case,[3] that,

$$\operatorname{supp} \mathcal{F}\theta \subseteq \operatorname{const} \Omega.$$

---

[3] See Lemma 6.3.1.

For large x, i.e., $\deg x \ll 0$, $Ef(x) = |x|^{1/2} f(0)$, hence we also require that $\Omega(0) = 0$. Then the image of M_Ω consists of functions on the adeles such that $f(0) = 0$. In fact, for f in the image, $Ef(x) = 0$ already for $\deg x < -\omega$. To assure that Ef is compactly supported, we also require that $\theta(0) = 0$. Indeed, by Riemann–Roch, $Ef(x) = E\mathcal{F}f(1/x)$, which vanishes for $-\deg x < -\theta$, so Ef is supported on $-\omega \le \deg x \le \theta$.

Then we might expect that

$$Z_{-\omega..\theta} - EC_{\mathcal{F}\theta}AM_\Omega$$

is a projection (onto a relatively small space), and hence a positive operator. Taking $h = k^* * k$, also U_h is a positive operator, and

$$\mathrm{tr}((Z_{-\omega..\theta} - EC_{\mathcal{F}\theta}AM_\Omega)U_h) \ge 0.$$

In Section 6.1, formula (6.2), we computed

$$\mathrm{tr}(Z_{-\omega..\theta}U_h) = (\omega + \theta + 1)h(1).$$

If we could establish the trace formula

$$\mathrm{tr}(EC_{\mathcal{F}\theta}AM_\Omega U_h) = (\omega + \theta - \deg \mathcal{K} - 1)h(1) + \sum_v \deg v \, W_v(h),$$

then Weil positivity would follow.

We see the following problems with this approach in the global and the semi-local case. First,

$$AEf(x) = \sum_{\alpha \in \ker |\cdot|/o^*} f(\alpha x)$$

is far from the identity operator: E does more than just make a function multiplicative. Second, we can arrange that $C_{\mathcal{F}\theta}$ and M_Ω commute and choose $\Omega(0) = \theta(0) = 0$, but neither cutoff commutes with E, and therefore $EC_{\mathcal{F}\theta}AM_\Omega$ is not a projection. We can see already in the semi-local computation how far from a projection it is. This will suggest to us some possible adjustments.

Alternatively, we can write $C_{\mathcal{F}\theta}$ as $\mathcal{F}M_\theta\mathcal{F}^*$ (see Section 6.5.3 below), and then apply the Riemann–Roch formula $E\mathcal{F} = \mathcal{I}_*E$ to find

$$EC_{\mathcal{F}\theta}AM_\Omega = \mathcal{I}_*EM_\theta\mathcal{F}^*AM_\Omega,$$

where $\mathcal{I}_*$ is the inversion on the space of coarse idele classes, which is easier to handle than the Fourier transform on the adeles. This is as much commutativity as remains in the global case between E and the convolution. This alternative is not available in the local and semi-local case, but it does not seem to give more than a minor simplification.

Since $EC_{\mathcal{F}\theta}AM_\Omega$ is not a projection into $Z_{-\omega..\theta}$, it is unclear if the above positivity argument works. This will depend on whether it approximates a projection closely enough.

6.5 The operators M, A, C, $\mathcal{F}^*$, and E

We will have a closer look at each operator involved in the cutoff on the adeles. The cutoff involves several function spaces: the shift acts on a function on $Z = \mathbb{A}^*/\ker|\cdot|$. The multiplication operator M_Ω lifts it to a function on the group of divisors $\mathbb{A}^*/\mathfrak{o}^*$. Next, the operator A changes the measure d^*a on this space to the "additive" measure $|a|\,d^*a$, and is an isomorphism of Hilbert spaces. Then $C_{\mathcal{F}\theta}$ lifts this to a function on $\mathbb{A}/\mathfrak{o}^*$, a function on the adeles that is $\mathfrak{o}^*$-invariant.

We study these operators in detail for the global adeles and ideles, and also for the semi-local case for a finite set S of valuations. The first nontrivial case is when S contains at least two valuations, and $S = \{v\}$ is the local case.

6.5.1 Restricting the support

A function on Z can be lifted to the group of divisors by

$$Lf(x\mathfrak{o}^*) = f(x\ker|\cdot|).$$

However, this allows ideles x of fixed degree with arbitrary large order of pole or zero: the set $\deg x = n$ inside the group of divisors is not compact.

To make the choice of x supported on a compact set of ideles, we multiply by a function with compact support in the adeles:

$$M_\Omega f(x\mathfrak{o}^*) = \Omega(x\mathfrak{o}^*)f(x\ker|\cdot|),$$

where Ω is a function on the adeles that is $\mathfrak{o}^*$-invariant. We have called this the infrared cutoff.

We first consider the choice $\Omega = \mathfrak{p}^{-\Omega}$, where Ω is a positive divisor of large degree (see the footnote on page 117). Then for an idele x, each component of x must satisfy $v(x) \geq -\Omega_v$. Moreover,

$$-\Omega_v \leq v(x) \leq \max\{0, (\deg\Omega + \deg x)/\deg v\},$$

for the following reason: the total degree of the polar part of x is

$$\sum_{v(x)<0} v(x) \deg v \geq -\deg \Omega.$$

If $v(x) > (\deg \Omega + \deg x)/\deg v$ for a valuation with $v(x) > 0$, then it would follow that $\deg x > \deg x$, a contradiction.

It follows that $v(x) = 0$ for $\deg v > \max\{0, \deg \Omega + \deg x\}$. So the choice of the divisor x is compactly supported, but this compact set varies with the degree of x. In other words, M_Ω is in general not compactly supported.

Exercise 6.5.1 Show that as a function $L^2(Z) \to L^2(\mathbb{A}^*/\mathfrak{o}^*, d^*a)$, M_Ω is not bounded.

Exercise 6.5.2 Show that $M_\Omega U(a) = U(a) M_{U(1/a)\Omega}$.

For the application of E (see Section 6.5.4) it is necessary that $f(0) = 0$ for a function in the image of M_Ω. This is ensured if $\Omega(0)$ vanishes. Therefore we choose

$$\Omega = \mathfrak{p}^{-\Omega} - \mathfrak{p}^0. \tag{6.11}$$

With this choice, $\Omega(x)$ vanishes in a neighborhood of zero in the adeles: an idele is close to zero if $v(x) \geq 0$ for all v and $v(x) \gg 0$ for many v. Then both $\mathfrak{p}^{-\Omega}(x)$ and $\mathfrak{p}^0(x) = 1$, so that $\Omega(x) = 0$.

In the S-local case, E_v does not satisfy Riemann–Roch, so it is unnecessary that $f(0) = 0$, and we can take $\Omega_S = \mathfrak{p}_S^{-\Omega}$ for some divisor Ω supported in S.

6.5.2 Making the measure additive

The operator $A \colon L^2(\mathbb{A}^*/\mathfrak{o}^*, d^*a) \to L^2(\mathbb{A}^*/\mathfrak{o}^*, |a|\, d^*a)$ is defined by

$$Af(x) = |x|^{-1/2} f(x).$$

Exercise 6.5.3 A is an isomorphism of Hilbert spaces.

Each multiplicative space, Z and $\mathbb{A}^*$ with the measure d^*a, has a shift operator

$$U(a)f(x) = f(x/a),$$

and on the additive spaces ($\mathbb{A}$ and $\mathbb{A}^*$ with the measure $|a|\, d^*a$), we have the dilation

$$D(a)f(x) = |a|^{-1/2} f(x/a). \tag{6.12}$$

Exercise 6.5.4 The shift $U(a)$ is unitary in $L^2(\mathbb{A}^*/\mathfrak{o}^*, d^*a)$, and $D(a)$ is unitary in $L^2(\mathbb{A}^*/\mathfrak{o}^*, |a|\, d^*a)$.

Exercise 6.5.5 The operator A intertwines the shift with the dilation:

$$AU(a) = D(a)A.$$

6.5.3 Smoothing the oscillations

We have already seen that $M_\Omega f$ vanishes in a neighborhood of zero in the adeles. To further control the small oscillations of a function on the ideles, we convolve it with a function $\mathcal{F}\theta$, where

$$\theta = \mathfrak{p}^{-\Theta} - \mathfrak{p}^{-\mathcal{K}}. \tag{6.13}$$

Then

$$\mathcal{F}\theta = q^{1-g+\theta}\mathfrak{p}^{-\mathcal{K}+\Theta} - q^{g-1}\mathfrak{p}^0,$$

which for $\Theta \gg 0$ is a function with a large peak in the small set $\pi^{\Theta-\mathcal{K}}\mathfrak{o}$.

We define

$$C_{\mathcal{F}\theta}f(x) = \int \mathcal{F}\theta(x-y)f(y)|y|\, d^*y. \tag{6.14}$$

This produces a function on the adeles that is $\mathfrak{o}^*$-invariant and locally constant at distances in $\pi^{\Theta-\mathcal{K}}\mathfrak{o}$. This is the ultraviolet cutoff.

The Fourier transform of this function is not hard to compute:

$$\mathcal{F}^{-1}C_{\mathcal{F}\theta}f(z) = \iint \chi(-xz)f(y)\mathcal{F}\theta(x-y)|y|\, d^*y\, dx$$

$$= \theta(z)\int \chi(-yz)f(y)|y|\, d^*y. \tag{6.15}$$

Since $\theta(0) = 0$ by (6.13), it follows in particular that

$$\mathcal{F}^{-1}C_{\mathcal{F}\theta}f(0) = 0. \tag{6.16}$$

Thus a function in the image satisfies the requirement that its Fourier transform vanishes at 0.

In formula (6.15), we see an expression that looks like an inverse Fourier transform, but on the multiplicative space $\mathbb{A}^*$ with the additive measure $|a|\, d^*a$. This leads us to introduce the operator

$$\mathcal{F}^*f(z) = \int \chi(-yz)f(y)|y|\, d^*y.$$

Both $\mathcal{F}^*$ and $C_{\mathcal{F}\theta}$ turn a function on the ideles into a function on the adeles. Further, we have shown that

$$C_{\mathcal{F}\theta} = \mathcal{F} M_\theta \mathcal{F}^*,$$

where M_θ is defined as in Section 6.5.1, but now acts on a function on the adeles.

Remark 6.5.6 The function $\mathcal{F}\theta(x-y)$ separates the idele y from the adele x. In the S-local case, the unit S-ideles $\mathfrak{o}_S^*$ are open in the adeles, but globally they have measure zero: the ideles form a dense null set inside the adeles. Thus the convolution $C_{\mathcal{F}\theta}$ (and also $\mathcal{F}^*$; both use the measure $|x|\,d^*x$) is not defined on the L^2-space of the adeles, but only on a space of test functions that are continuous and locally constant, so that the extrapolation to the adeles is meaningful (see also Remark 3.1.7).

The operators $\mathcal{F}^*$ and $C_{\mathcal{F}\theta}$ are not continuous in the L^2-norms of functions on $\mathbb{A}^*$ with the additive measure to functions on $\mathbb{A}$. Still, for $f = \mathfrak{p}^D$ we have

$$\mathcal{F}^* \mathfrak{p}^D(x) = \prod_v \int_{K_v^*} \chi_v(-x_v y) \mathfrak{p}_v^{D_v}(y)|y|_v \, d_v^* y,$$

hence we find $\mathcal{F}^* \mathfrak{p}^D = q^{1-g-\deg D} \mathfrak{p}^{-\mathcal{K}-D}$, just as with the ordinary Fourier transform, cf. (3.2).

Exercise 6.5.7 Show that $\mathcal{F}^* D(a) = D(1/a)\mathcal{F}^*$.

Since they act on incompatible spaces, the operators M_Ω and $C_{\mathcal{F}\theta}$ cannot commute. However, if $\operatorname{supp} \mathcal{F}\theta \subseteq \operatorname{const} \Omega$ then we can still prove that $C_{\mathcal{F}\theta} A M_\Omega$ maps to functions supported within $\operatorname{supp}\Omega$. We have

$$C_{\mathcal{F}\theta} A M_\Omega f(x) = \int \mathcal{F}\theta(x-y)\Omega(y)f(y)|y|^{1/2}\, d^*y,$$

hence $x-y$ is restricted by the support of $\mathcal{F}\theta$, so that $x-y \in \operatorname{const}\Omega$. It follows that $\Omega(y) = \Omega(x)$, and we conclude that

$$C_{\mathcal{F}\theta} A M_\Omega f(x) = \Omega(x)\int \mathcal{F}\theta(x-y)f(y)|y|^{1/2}\, d^*y.$$

This is formally $M_\Omega C_{\mathcal{F}\theta} A f(x)$, and vanishes unless $\Omega(x) \neq 0$. In particular, by (6.11), it follows that

$$C_{\mathcal{F}\theta} A M_\Omega f(0) = 0.$$

We already saw in (6.16) above that a function in the image of $C_{\mathcal{F}\theta}$ also satisfies the requirement that its Fourier transform vanishes at zero.

6.5.4 Restricting to the coarse idele classes

We have verified that $f(0) = \mathcal{F}f(0) = 0$ and that f is locally constant and compactly supported for a function in the image of $C_{\mathcal{F}\theta}AM_\Omega$. For such a function on the adeles, $Ef(x)$ is compactly supported on Z.

Exercise 6.5.8 $E: L^2(\mathbb{A}, dx) \to L^2(Z)$ is an unbounded operator. In the S-local case, E is bounded.

By Riemann–Roch, $EC_{\mathcal{F}\theta}AM_\Omega = \mathcal{I}_* EM_\theta \mathcal{F}^* AM_\Omega$ in the global case.

Exercise 6.5.9 Show that the average restriction E intertwines the shift on Z with the dilation on $\mathbb{A}/\mathfrak{o}^*$,

$$U(a)E = ED(a).$$

6.6 Semi-local and global trace

Recall the ring of S-adeles from Section 3.5. The cutoff on this space of the combined shift by the group of S-degrees (coarse S-idele classes),

$$Z_S = \mathbb{A}_S^* / \ker |\cdot|_S,$$

is accomplished in three steps, $C_{\mathcal{F}_S\theta}A_S M_\Omega$. The two cutoffs commute provided $\Theta + \Omega \geq \mathcal{K}_S$, where $\mathcal{K}_S$ is the canonical S-divisor

$$\mathcal{K}_S = \sum_{v \in S} k_v v.$$

The local case is when $S = \{v\}$, and the global case is when S contains all valuations. There is, however, one important difference between the S-local and the global case: in the global case, we need that Ω and θ vanish at 0, because then the average restriction maps to a space of functions that are compactly supported in Z, or equivalently, the Mellin transform $\mathcal{M}Ef$ has no poles.

A function on $\mathbb{A}_S/\mathfrak{o}_S^*$ is mapped to an idele class function by the average restriction

$$E_S f(m) = \sqrt{|m|_S} \sum_{\alpha \in \ker |\cdot|_S/\mathfrak{o}_S^*} f(m\alpha).$$

We have seen in Section 3.5 that it defines the (shifted) zeta function as a Mellin transform.

We compute the kernel of the combined shift $U_h = \sum_{n \in Z_S} h(n) U_S(n)$, cut off on the additive space of S-adeles. Writing d for the general element of $\mathbb{A}_S^* / \mathfrak{o}_S^*$ (an S-divisor), we have, for a function f on Z_S,

$$A_S M_\Omega U_h f(d) = \Omega(d) \sum_{n \in Z_S} h(n) f(d/n) |d|_S^{-1/2}.$$

Convolving with $\mathcal{F}_S \theta$, we obtain

$$C_{\mathcal{F}_S \theta} A_S M_\Omega U_h f(x) = \sum_n h(n) \sum_d \mathcal{F}_S \theta(x - d) \Omega(d) f(d/n) |d|_S^{1/2}.$$

We take the average over $\mathfrak{o}_S^*$ and substitute $\varepsilon n d$ for d to obtain

$$E_S C_{\mathcal{F}_S \theta} A_S M_\Omega U_h f(m)$$
$$= \sum_d f(d) \sum_n h(n) \sum_\alpha \int \mathcal{F}_S \theta(m\alpha - \varepsilon n d) \Omega(nd) |mnd|_S^{1/2} d_S^* \varepsilon,$$

where α runs over the group $\ker |\cdot|_S / \mathfrak{o}_S^*$ of divisors of vanishing degree.

Changing the measure $d_S^* \varepsilon$ to $\kappa_+^* d_S \varepsilon$, where $\kappa_+^* = \prod_{v \in S} \kappa_+^*(v)$, we see the Fourier transform[4] $\mathcal{F}_S(\mathfrak{o}_S^*)$. Substituting $x/(nd) = x/nd$ for x yields the following formula for $E_S C_{\mathcal{F}_S \theta} A_S M_\Omega U_h f(m)$:[5]

$$\kappa_+^* \sum_d f(d) \sum_n h(n) \sum_\alpha \int \chi_S(xm\alpha/nd) \mathcal{F}_S(\mathfrak{o}_S^*)(x)$$
$$\times \theta(x/nd) \Omega(nd) |nd/m|_S^{-1/2} d_S x.$$

To see the kernel, we write $d = l\beta$, where l runs over all degrees $l \in Z_S$ and β runs over all divisors of degree zero, $\beta \in \ker |\cdot|_S / \mathfrak{o}_S^*$. Replacing α by $\alpha\beta$, the kernel is found as

$$k(m, l) = \kappa_+^* \sum_n h(n) |nl/m|_S^{-1/2} \sum_{\alpha, \beta} \int \chi_S(xm\alpha/nl) \mathcal{F}_S(\mathfrak{o}_S^*)(x)$$
$$\times \theta(x/nl\beta) \Omega(nl\beta) d_S x.$$

The trace is the sum over the diagonal $m = l$. Noting that $nm\beta$ runs over all divisors, we write d for $nm\beta$, to find the trace

$$\sum_m k(m, m)$$
$$= \kappa_+^* \sum_n h(n) |n|_S^{-1/2} \sum_\alpha \int \chi_S(x\alpha/n) \mathcal{F}_S(\mathfrak{o}_S^*)(x) \sum_d \theta(x/d) \Omega(d) d_S x.$$

---

[4] In the global case, we use the "Fourier transform" $\mathcal{F}^*$ on the ideles, defined in Section 6.5.3.

[5] To improve readability, we write $xm\alpha/nd$ for $xm\alpha/(nd)$, etc.

Replacing $\kappa_+^* d_S x$ by $|x|_S d_S^* x$ and then writing the integral over x as a sum over all divisors e and an integral over $\mathfrak{o}_S^*$ with the measure $d_S^* \varepsilon$, we obtain

$$\sum_n h(n)|n|_S^{-1/2} \sum_{\alpha,e} \int \chi_S(e\varepsilon\alpha/n)\mathcal{F}_S(\mathfrak{o}_S^*)(e)$$

$$\times \sum_d \theta(e/d)\Omega(d)h(n)|n|_S^{-1/2}|e|_S\, d_S^*\varepsilon.$$

The sum over divisors d is easy to compute:

$$\sum_d \theta(e/d)\Omega(d) = \prod_{v\in S}\left(\Theta_v + \Omega_v + e_v + 1\right),$$

since $-\Omega \leq d \leq \Theta + e$, hence $-\Omega_v \leq d_v \leq \Theta_v + e_v$ for each $v \in S$. This also holds in the global case since $\Theta_v + \Omega_v + e_v$ vanishes for all but finitely many valuations v. Moreover, each factor is positive since $\Theta + \Omega \geq \mathcal{K}$ and e is in the support of $\mathcal{F}_S \mathfrak{o}_S^*$:

$$\mathcal{F}_S(\mathfrak{o}_S^*) = q^{-\deg \mathcal{K}_S/2} \prod_{v\in S}\left(\mathfrak{p}_v^{-k_v} - q_v^{-1}\mathfrak{p}_v^{-k_v-1}\right),$$

so that $e_v \geq -k_v - 1$.

Writing $\mu(D)$ for the Möbius function of a divisor D:

$$\mu(D) = \prod_v (-1)^{D_v}$$

if $D_v \in \{0,1\}$ for all v and otherwise $\mu(D)$ vanishes, we may write this as

$$\mathcal{F}_S(\mathfrak{o}_S^*) = q^{-\deg \mathcal{K}_S/2} \sum_{d|S} \mu(d)|d|_S \mathfrak{p}^{-\mathcal{K}_S-d}.$$

However, this does not help much: we see already that the product is in general much larger than $\theta + \omega + 1 + \deg e$, hence we get a result that is very different from Sections 6.2 and 6.3. The abstract theory from Section 6.2 does not apply because $\mathfrak{p}^{-\Omega}$ does not approximate $\delta_{\deg x \geq -\omega}$ closely enough, and we will not recover the explicit formula. This discrepancy occurs as soon as S contains at least two valuations.

It may be possible to adjust the functions Ω and θ to remedy this situation. In the next section, we derive an Euler product for the kernel that may help to investigate alternative choices for Ω and θ.

6.7 The kernel on the analytic space

Combining with the isomorphism $\mathcal{M}\colon \mathcal{Z} \to \mathcal{H}$, we have the operator

$$\mathcal{M}EC_{\mathcal{F}\theta}AM_{\Omega}U_{h}\mathcal{M}^{-1}.$$

We compute the kernel of this operator. The computation will be the same in the global and the S-local case, with one important difference in the global case that we will discuss.

By (6.1), the combined shift is a multiplication operator:

$$U_{h}\mathcal{M}^{-1}f(n) = \mathcal{M}^{-1}(hf)(n),$$

where we write $h(s) = \sum_{n} h(n)|n|^{s}$ for the Mellin transform of h. Thus we find

$$\begin{aligned}
\big(&\mathcal{M}EC_{\mathcal{F}\theta}AM_{\Omega}U_{h}\mathcal{M}^{-1}f\big)(s) \\
&= \sum_{n \in \mathcal{Z}} \big(EC_{\mathcal{F}\theta}AM_{\Omega}\mathcal{M}^{-1}(hf)\big)(n)|n|^{s} \\
&= \sum_{n} \sum_{\alpha \in \ker|\cdot|/\mathfrak{o}^{*}} \big(C_{\mathcal{F}\theta}AM_{\Omega}\mathcal{M}^{-1}(hf)\big)(n\alpha)\,|n|^{1/2+s} .
\end{aligned}$$

The sum over all degrees $n \in \mathbb{A}^{*}/\ker|\cdot|$ and all divisors of vanishing degree α can be combined into a sum over all divisors d, with corresponding idele $\pi^{d} \in \mathbb{A}^{*}/\mathfrak{o}^{*}$, to obtain

$$\begin{aligned}
\sum_{d} &\big|\pi^{d}\big|^{1/2+s}\big(C_{\mathcal{F}\theta}AM_{\Omega}\mathcal{M}^{-1}(hf)\big)\big(\pi^{d}\big) \\
&= \sum_{d} \big|\pi^{d}\big|^{1/2+s} \int_{\mathbb{A}^{*}} \mathcal{F}\theta\big(\pi^{d} - x\big)\big(AM_{\Omega}\mathcal{M}^{-1}(hf)\big)(x)\,|x|\,d^{*}x.
\end{aligned}$$

The choice of the idele π^{d} is up to an element of $\mathfrak{o}^{*}$ but the integral is independent of this choice.

Applying A and M_{Ω} and using the inverse Mellin transform of (4.2), we obtain

$$\begin{aligned}
\sum_{d} &\big|\pi^{d}\big|^{1/2+s} \int_{\mathbb{A}^{*}} \mathcal{F}\theta\big(\pi^{d} - x\big)\Omega(x)\mathcal{M}^{-1}(hf)(x)\,|x|^{1/2}\,d^{*}x \\
&= \sum_{d} \big|\pi^{d}\big|^{1/2+s} \int_{\mathbb{A}^{*}} \mathcal{F}\theta\big(\pi^{d} - x\big)\Omega(x) \fint (hf)(t)\,|x|^{-t}\,dt\,|x|^{1/2}\,d^{*}x \\
&= \fint f(t)h(t) \sum_{d} \big|\pi^{d}\big|^{1/2+s} \int_{\mathbb{A}^{*}} \mathcal{F}\theta\big(\pi^{d} - x\big)\Omega(x)\,|x|^{1/2-t}\,d^{*}x\,dt.
\end{aligned}$$

Initially, the integral over x has the same abscissa of convergence as the zeta function, $\mathrm{Re}\, t < -1/2$.[6] We shall see below that the integral actually converges for all t since we are taking a function for Ω that vanishes in a neighborhood of 0. Likewise, the sum over d converges for $\mathrm{Re}\, s > 1/2$, but given our choice of θ below, it converges for all s.

The kernel of $\mathcal{M}EC_{\mathcal{F}\theta}AM_{\Omega}U_h\mathcal{M}^{-1}$ is found as the following function of period $2\pi i/\log q$ in the two variables s and t:

$$k(s,t) = h(t) \sum_d |\pi^d|^{1/2+s} \int_{\mathbb{A}^*} \mathcal{F}\theta(\pi^d - x)\Omega(x)|x|^{1/2-t}\, d^*x.$$

We can conveniently compute this function by writing it as an Euler product: d^*x is a product measure on the ideles, the function $\Omega(x)$ is also a product $\prod_v \Omega_v(x_v)$ and so is $\mathcal{F}\theta$, hence

$$\sum_d |\pi^d|^{1/2+s} \int \mathcal{F}\theta(\pi^d - x)\Omega(x)|x|^{1/2-t}\, d^*x$$

$$= \prod_v \sum_{d=-\infty}^{\infty} |\pi_v^d|_v^{1/2+s} \int_{K_v^*} \mathcal{F}_v\theta_v(\pi_v^d - x)\Omega_v(x)|x|_v^{1/2-t}\, d_v^*x.$$

The factors are computed in the following two lemmas:

Lemma 6.7.1 *For $\Omega = \mathfrak{p}_v^{-\omega}$, $\theta = \mathfrak{p}_v^{-\theta}$ and $\mathrm{Re}\, s > -1/2$, $\mathrm{Re}\, t < 1/2$,*

$$\sum_{d=-\infty}^{\infty} |\pi_v^d|_v^{1/2+s} \int_{K_v^*} \mathcal{F}_v\theta(\pi_v^d - x)\Omega(x)|x|_v^{1/2-t}\, d_v^*x$$

$$= \frac{q_v^{k_v/2} q_v^{(\theta - k_v)(t-s)}}{\left(1 - q_v^{-1/2+t}\right)\left(1 - q_v^{-1/2-s}\right)} + \frac{q_v^{k_v/2}}{1 - q_v^{-1}} \sum_{k=-\omega}^{\theta - k_v - 1} q_v^{k(t-s)}.$$

Proof For fixed d, the integral is supported on $v(x) \geq -\omega$,

$$\int_{K_v^*} \mathcal{F}_v\theta(\pi_v^d - x)\Omega(x)|x|_v^{1/2-t}\, d_v^*x$$

$$= \sum_{k=-\omega}^{\infty} q_v^{-k(1/2-t)} \int_{\mathfrak{o}_v^*} \mathcal{F}_v\theta(\pi_v^d - \pi_v^k x)\, d_v^*x.$$

Now $\mathcal{F}_v\theta = q_v^{\theta - k_v/2}\mathfrak{p}_v^{\theta - k_v}$, hence the function $\mathcal{F}_v\theta(\pi_v^d - \pi_v^k x)$ vanishes unless $d, k \geq \theta - k_v$ or $d = k$ and $k < \theta - k_v$. For the sum over d, we

---
[6] We can move the line of integration to $\mathrm{Re}\, t = \tau < -1/2$ provided $h(t)$ and $f(t)$ are defined and holomorphic in the strip $\{t: \tau \leq \mathrm{Re}\, t \leq 0\}$.

obtain

$$\sum_{d=-\infty}^{\infty} q_v^{-d(1/2+s)} \int_{K_v^*} \mathcal{F}_v \theta\left(\pi_v^d - x\right) \Omega(x) |x|_v^{1/2-t} d_v^* x$$

$$= \sum_{k=\theta-k_v}^{\infty} q_v^{-k(1/2-t)} \sum_{d=\theta-k_v}^{\infty} q_v^{-d(1/2+s)} \int_{\mathfrak{o}_v^*} \mathcal{F}_v \theta\left(\pi_v^d - \pi_v^k \varepsilon\right) d_v^* \varepsilon$$

$$+ \sum_{k=-\omega}^{\theta-k_v-1} q_v^{-k(1/2-t)} q_v^{-k(1/2+s)} \int_{\mathfrak{o}_v^*} \mathcal{F}_v \theta\left(\pi_v^k(1-x)\right) d_v^* x.$$

The integral over ε equals $q_v^{\theta-k_v/2}$, hence we find the value

$$q_v^{\theta-k_v/2} \frac{q_v^{(k_v-\theta)(1/2-t)}}{1-q_v^{-1/2+t}} \cdot \frac{q_v^{(k_v-\theta)(1/2+s)}}{1-q_v^{-1/2-s}}$$

for the double sum over k and d. The second sum (over k in the third line) is found by the next lemma and yields the formula as stated. $\qquad\square$

Lemma 6.7.2 *For* $k \leq \theta - k_v - 1$,

$$\int_{\mathfrak{o}_v^*} \mathcal{F}_v \theta\left(\pi_v^k(1-x)\right) d_v^* x = \frac{q_v^{k_v/2+k}}{1-q_v^{-1}}.$$

Proof Changing to the additive measure, the integral equals

$$\kappa_+^* \int_{\mathfrak{o}_v - \pi_v \mathfrak{o}_v} \mathcal{F}_v \theta\left(\pi_v^k(1-x)\right) d_v x.$$

The integral over $\pi_v \mathfrak{o}_v$ vanishes since for $v(x) \geq 1$, $\pi_v^k(1-x)$ has valuation k, which is outside of the support of $\mathcal{F}_v \theta$. Replacing $1-x$ by x, we find

$$\kappa_+^* \int_{\mathfrak{o}_v} \mathcal{F}_v \theta\left(\pi_v^k x\right) d_v x = \kappa_+^* q_v^{\theta-k_v/2} \int_{\mathfrak{o}_v} \mathfrak{p}_v^{\theta-k_v}\left(\pi_v^k x\right) d_v x.$$

This turns out to be an integral over the set $v(x) \geq \theta - k_v - k \geq 1$, with value $\kappa_+^* q_v^{\theta-k_v/2} q_v^{k-\theta+k_v/2}$. $\qquad\square$

With Lemma 6.7.1, using the Euler product for ζ_C and the degree of the canonical divisor $\sum_v k_v \deg v = 2g - 2$, we can write the kernel as

$$k(s,t) = h(t)\Lambda_C(-t)\Lambda_C(s)q^{(\theta-g+1)(t-s)} \times$$

$$\times \prod_v \left(1 + \frac{\left(1-q_v^{-1/2+t}\right)\left(1-q_v^{-1/2-s}\right)}{1-q_v^{-1}} \sum_{k=1}^{\Omega_v+\Theta_v-k_v} q_v^{k(s-t)}\right).$$

As in (6.7), we assume that $\Omega + \Theta \geq \mathcal{K}$ (see also Section 6.5.3), so that $\Omega_v + \Theta_v \geq k_v$ for every valuation. For only finitely many valuations, the inequality holds, and for every other valuation, the factor in the product is trivial.

We have computed the kernel for $\Omega = \mathfrak{p}^{\Omega}$ and $\theta = \mathfrak{p}^{\Theta}$. This suffices for the local and S-local cases. But for the global case, it is necessary that the functions Ω and θ vanish at 0. We will show that these vanishing conditions cancel the poles of $\Lambda_{\mathcal{C}}(s)\Lambda_{\mathcal{C}}(-t)$ at $s = \pm 1/2$, $t = \pm 1/2$, and then periodically modulo $2\pi i / \log q$.

For $t = 1/2$, the factors in the product are trivial, and hence $k(s, 1/2)$ does not depend on Ω. To indicate the dependence on Ω and θ, let us write

$$k(s, t) = k_{\Omega,\Theta}(s, t).$$

We see that $k_{\Omega,\Theta}(s, 1/2) - k_{0,\Theta}(s, 1/2)$ vanishes, and it follows that the function

$$k_{\Omega,\Theta} - k_{0,\Theta}$$

has the factor $1 - q^{-1/2+t}$. Similarly, $k_{\Omega,\Theta}(-1/2, t)$ does not depend on Ω, hence the above difference has the factor $1 - q^{-1/2-s}$ as well. This cancels the poles at $t = 1/2$ and at $s = -1/2$.

With a little more computation, we find that $k_{\Omega,\Theta}(s, -1/2)$ does not depend on Θ, and neither does $k_{\Omega,\Theta}(1/2, t)$. We conclude that $k_{\Omega,\Theta} - k_{\Omega,\mathcal{K}}$ has the factor $\left(1 - q^{-1/2-t}\right)\left(1 - q^{-1/2+s}\right)$.

We find that the kernel

$$k_{\Omega,\Theta} - k_{\Omega,\mathcal{K}} - k_{0,\Theta} + k_{0,\mathcal{K}}$$

has no poles and is a trigonometric polynomial with finitely many positive and negative powers of q^s and q^t.

We are not able to compute this kernel more explicitly, or understand it better. For $s = t$, $k(s, t)$ simplifies significantly (see also Exercise 6.3.4):

Exercise 6.7.3 Find the trace $\int k(s, s)\, ds$. Locally, for $S = \{v\}$, we recover the local trace formula.

7

Epilogue

This book has been aimed at the first step in the following two-step program for the Riemann hypothesis:

(i) Work out Connes' approach in the geometric case of a curve over a finite field.

(ii) Translate this approach to a number field, the arithmetic case.

The second step naturally divides into two steps:

(a) Translate the local theory (archimedean and nonarchimedean case).

(b) Translate the global theory.

There is no difficulty in the translation of Connes' approach to the p-adic completions of a number field (step (iia) in the nonarchimedean case). This can already be found in Haran's work [Har3] and gives the Weil distribution as in Section 4.3. It is remarkable that also the archimedean local trace formula can already be found in the work of Weil, Haran, and Connes. The global theory, on the other hand, is incomplete, both in Chapter 6 of this book, and in the work of Connes and Haran.

One may wonder how much of the work towards a proof of the Riemann hypothesis will be accomplished with step (i). Is the geometric case essentially equivalent to the Riemann hypothesis? Or is the geometric case in a fundamental way simpler? One is reminded of the abc-conjecture, where the solution in the geometric case is fundamentally simpler and does not seem to give much insight for the arithmetic case. We are optimistic and hope that the completion of step (i) is a major step towards the Riemann hypothesis, and that, apart from overcoming technical difficulties, not more is needed.

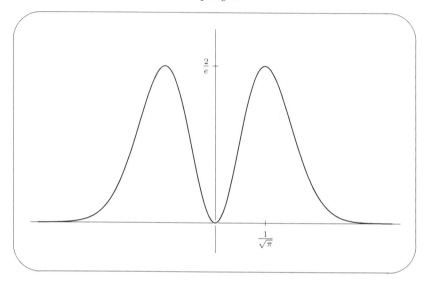

Figure 7.1 The shifted Gaussian.

7.1 Archimedean translation

We have already seen one example of a translation to the archimedean case in Section 6.2: the function $\mathfrak{p}_v^0 - q_v \mathfrak{p}_v^1$ corresponds to the difference between the Gaussian $e^{-\pi x^2}$ and the dilated Gaussian,

$$\frac{e^{-\pi x^2} - u e^{-\pi x^2 u^2}}{u - 1}.$$

The group of coarse idele classes of $\mathbb{Q}$ is

$$\mathbb{A}^* / \ker |\cdot| \cong \mathbb{R}^+.$$

This group allows infinitesimal dilations: for $u \to 1$, we find the derivative $(1 - 2\pi x^2)e^{-\pi x^2}$ (see Figure 6.1 on page 113).

Likewise, shifting the Gaussian multiplicatively, we consider

$$\frac{e^{-\pi x^2} - e^{-\pi x^2 v^2}}{v - 1},$$

which gives the infinitesimal multiplicative shift $2\pi x^2 e^{-\pi x^2}$ of Figure 7.1.

The combination is like a second derivative (see Remark 3.3.2 and Figure 7.2): dilating by u and shifting by v and letting $u, v \to 1$, we find the function

$$G(x) = 2\pi x^2 (3 - 2\pi x^2) e^{-\pi x^2},$$

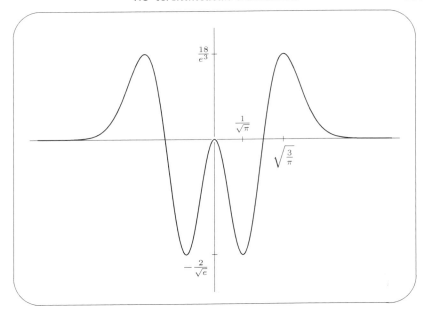

Figure 7.2 The shifted and dilated Gaussian.

which is the analogue of $\mathfrak{p}_v^0 - (q+1)\mathfrak{p}_v^1 + q\mathfrak{p}_v^2$. Not surprisingly, the Mellin transform on $\mathbb{R}$ (see Remark 4.1.1) of the corresponding theta function,

$$\sum_{n=-\infty}^{\infty} G(nx),$$

gives the completed Riemann zeta function with its poles cancelled,

$$s(s-1)\zeta_{\mathbb{Z}}(s),$$

where $\zeta_{\mathbb{Z}}$ is given in the introduction, equation (1).

   This is the starting point from where the archimedean theory takes off. All definitions and theorems in this book have their counterpart for $\mathbb{Q}$, except in Chapter 5, where the two-dimensional geometric space $\mathcal{C} \times \mathcal{C}$ is essential, for which we do not know an arithmetic counterpart.

   To point out just one more problem that would have to be addressed: convolution and multiplication never commute; the uncertainty principle already occurs at the infinitesimal level. All that remains is approximate commutativity. See [Li] for an approach that replaces the Fourier transform by another operator. Also Haran [Har4] has uncovered a remarkable and beautiful structure based on monoids and generalized rings, inspired by the work of Christophe Soulé [So]. The real prime in Haran's work is naturally recovered in the structure of the orthogonal group. See also [CoCo1, CoCo4] for a related point of view.

Our optimistic expectation is that such considerations and analogies, as were already known to Tate and Weil, will provide a complete translation to the archimedean completions of a number field, which together with the p-adic theories will combine to create the global Weil explicit formula and Weil positivity. In other words, a complete proof in the geometric case will translate to a proof of the Riemann hypothesis. A translation to the arithmetic case will not be without difficulty of course, but the difficulties are expected to be of a technical nature. Thus we expect that the geometric case is essentially as hard as the Riemann hypothesis itself. This would explain why we in this book, and others before us, have not succeeded.

7.2 Global translation

In Section 5.1, we have seen the core reason that the Riemann hypothesis holds in the geometric case, namely, inequality (5.1) between a projection in the space of adeles classes and the corresponding projection in the idele class space. In the arithmetic case, we expect that this inequality becomes an approximate or asymptotic inequality, because the analogue of $\mathfrak{p}^0$ contains as an archimedean component the Gaussian, which is not compactly supported but only of fast decay, and hence only approximately a projection.

The cutoff of Sections 6.2 and 6.4 needs to satisfy the following conditions:

(i) The functions Ω and θ are test functions on the adeles (locally constant and compactly supported).

(ii) They are $\ker |\cdot|$-invariant.

(iii) The corresponding operators M_Ω and C_θ are commuting projections.

These conditions can be satisfied in the local case, as we have seen in Section 6.3 and the beginning of Section 6.4. Globally, we need in addition that Ω and θ vanish at 0. This is no problem, but the first and second conditions are incompatible on the adeles. In Sections 6.4–6.6, we have only been able to satisfy the second requirement in a rough approximation.

We have seen many examples of trace computations in Chapter 6. The noncommutative point of view is that a trace is an integral over a noncommutative space, which suggests that, globally, Weil's explicit formula should be viewed as an integral over the space of adele classes.[1] A trace has the commuting property that $\text{tr}(AB) = \text{tr}(BA)$. If a trace does not exist, one may instead attempt to relax this property and construct KMS states, which satisfy this same property, up to a twist by an automorphism. See [CMR, CoCo3, Lac-vF1, Lac-vF2] for more information about KMS states in number theory.

7.3 The space of adele classes

It seems that, to make progress, a better understanding of the adele class space and its connection with the idele class group via the restriction is needed. Connes, Consani, and Marcolli have focused on this in their work on the adele class space $\mathbb{A}/K^*$. In our simplified approach, where we are interested in the Dedekind zeta function but not in L-series with a nontrivial character, we aim instead for a better understanding of $\mathbb{A}/\ker|\cdot|$.

Note that the subset $\mathbb{A}^*$ is dense in the adeles. Given an adele, there are three cases to consider to find its class in $\mathbb{A}/\ker|\cdot|$. First, if the adele is an idele then its class is found inside the coarse idele class group as its degree. If the adele is not an idele, then its norm vanishes for one or both of the following reasons: a component of the adele vanishes, or infinitely many components have positive valuation (the adele has infinitely many zeros). This corresponds to the two alternatives (formulated for $\mathbb{Q}$) in [CoCo2, Theorem 7.5]: $\mathbb{A}/\mathbb{Q}^*$ is a projective space (as explained in *loc. cit.*) and a morphism from $\mathbb{Z}[T]$ is determined by an idempotent associated with an idele, which is in turn determined by the zero divisor of the adele, or such a morphism is determined by a unique adele (unique in $\mathbb{A}/\mathbb{Q}^*$), which gives the evaluation map as in *loc. cit.*

Connes and Consani have uncovered a remarkably rich structure of the adele class space, which turns out to be related to the theory of hyperrings of Krasner. The adele class space is isomorphic to the fundamental group of the maximal abelian cover of the curve associated with the function field K, see [CoCo2, Theorem 7.12]. In the introduction on page 6, we have seen that the maximal abelian extension of K splits

---

[1] Provided that Haran's discovery that Weil's explicit formula can be written as a trace points in the right direction.

naturally into an arithmetic component of the maximal constant field extension, and a geometric component of abelian covers. See also [CCM] for more fascinating discoveries.

References

[Art1] Artin, E., Quadratische Körper im Gebiete der höheren Kongruenzen I, II, *Math. Zeitschr.* **19** (1924).

[Art2] Artin, E., *Algebraic Numbers and Algebraic Functions*, Amer. Math. Soc., Providence, RI, 2000. (Reprinted from the 1967 edition.)

[Bak] Baker, A., *A Comprehensive Course in Number Theory,* Cambridge University Press, 2012.

[Bom1] Bombieri, E., Counting points on curves over finite fields, *Seminaire Bourbaki* **430** (1973).

[Bom2] Bombieri, E., Hilbert's 8th problem: an analogue, in: *Mathematical Developments Arising from Hilbert Problems*, Proc. Symposia in Pure Mathematics **XXVIII**, Amer. Math. Soc., Providence, RI, 1976, 269–274.

[Bor] Borger, J., Λ-Rings and the field with one element, arxiv.org, 0906.3146.pdf.

[BC] Bost, J.-B., Connes, A., Hecke algebras, type III factors and phase transitions with spontaneous symmetry breaking in number theory, *Sel. Math., New Ser.* **1** (1995), 411–457.

[Conn1] Connes, A., Trace formula in noncommutative geometry and the zeros of the Riemann zeta function, *Sel. Math., New Ser.* **5** (1999), 29–106.

[Conn2] Connes, A., Noncommutative geometry and the Riemann zeta function, preprint, 2000.

[CoCo1] Connes, A., Consani, K., Schemes over $\mathbb{F}_1$ and zeta functions, *Compositio Mathematica* **146** (6) (2010), 1383–1415.

[CoCo2] Connes, A., Consani, K., The hyperring of adèle classes, *J. Number Theory* **131** (2011), 159–194.

[CoCo3] Connes, A., Consani, K., On the arithmetic of the BC-system, arXiv:1103.4672 [math.QA], 2011.

[CoCo4] Connes, A., Consani, K., The universal thickening of the field of real numbers, preprint, 2012, www.alainconnes.org/docs/witt_infty.pdf.

[CCM] Connes, A., Consani, K., Marcolli, M., The Weil proof and the geometry of the adeles class space, arXiv:math/0703392v1, 2007.

[CMR] Connes, A., Marcolli, M., Ramachandran, N., KMS states and complex multiplication, www.alainconnes.org/docs/cmr0.pdf and CMR1.pdf, 2004.

[Conr] Conrey, J. B., The Riemann hypothesis, *Notices Amer. Math. Soc.*, March 2003, 341–353.

[dV1] de la Vallée-Poussin, C.-J., Recherches analytiques sur la théorie des nombres; Première partie: La fonction $\zeta(s)$ de Riemann et les nombres premiers en général, *Ann. Soc. Sci. Bruxelles Sér. I* **20** (1896), 183–256.

[dV2] de la Vallée-Poussin, C.-J., Sur la fonction $\zeta(s)$ de Riemann et le nombre des nombres premiers inférieurs à une limite donnée, *Mém. Couronnés et Autres Mém. Publ. Acad. Roy. Sci., des Lettres, Beaux-Arts Belg.* **59** (1899–1900).

[Del] Deligne, P., La conjecture de Weil I, *Publ. Math. Inst. Hautes Études Sci.* **48** (1974), 273–308.

[Den1] Deninger, C., Lefschetz trace formulas and explicit formulas in analytic number theory, *J. reine angew. Math.* **441** (1993), 1–15.

[Den2] Deninger, C., Evidence for a cohomological approach to analytic number theory, in: *Proc. First European Congress of Mathematics* (A. Joseph *et al.*, eds.) **I**, Paris, July 1992, Birkhäuser-Verlag, Basel, 1994, 491–510.

[Ed] Edwards, H. M., *Riemann's Zeta Function*, Dover Books, New York, 2001.

[Ei] Eichler, M., *Introduction to the Theory of Algebraic Numbers and Functions*, Pure and Applied Mathematics **23**, Academic Press, New York, 1966.

[FreiK] Freitag, E., Kiehl, R., *Etale Cohomology and the Weil Conjecture*, Springer-Verlag, Berlin, Heidelberg, New York, 1988.

[GVF] Gracia-Bondia, J. M., Varilly J. C., Figueroa, H., *Elements of Noncommutative Geometry*, Birkhäuser Advanced Texts, Boston, MA, 2001.

[Had] Hadamard, J., Sur la distribution des zéros de la fonction $\zeta(s)$ et ses conséquences arithmétiques, *Bull. Soc. Math. France* **24** (1896), 199–220.

[Har1] Haran, S., Index theory, potential theory, and the Riemann hypothesis, in: *L-Functions and Arithmetic* (J. Coates, M. J. Taylor, eds.), London Mathematical Society Lecture Notes Series **153**, 1989, 257–270.

[Har2] Haran, S., On Riemann's zeta function, in: *Dynamical, Spectral, and Arithmetic Zeta Functions* (M. L. Lapidus, M. van Frankenhuijsen, eds.), Contemporary Mathematics **290**, Amer. Math. Soc., Providence, RI, 2001, 93–112.

[Har3] Haran, S., *The Mysteries of the Real Prime*, London Mathematical Society monographs, New Series **25**, Clarendon Press, Oxford, 2001.

[Har4] Haran, S., Non-additive geometry, *Compositio Mathematica* **143** (2007), 618–688.

[Hart] Hartshorne, R., *Algebraic Geometry*, Graduate Texts in Mathematics, Springer-Verlag, 1977.

[Has1] Hasse, H., Ueber Kongruenzzetafunktionen, *S. Ber. Preuß. Ak. Wiss.* 1934, p. 250.

[Has2] Hasse, H., *Zahlentheorie*, Berlin, 1949.

[Hay] Hayman, W. K., *Meromorphic Functions*, Oxford University Press, London, 1975.

[HiRoe] Higson, N., Roe, J., *Surveys in Noncommutative Geometry*, Clay Mathematics Proceedings **6**, Amer. Math. Soc., Providence, RI, 2006.

[In] Ingham, A. E., *The Distribution of Prime Numbers*, Cambridge University Press, 1932.

[Iw] Iwasawa, K., On the rings of valuation vectors, *Annals of Math., 2nd Ser.* **57** (1953), 331–356.

[K1] Katz, N. M., An overview of Deligne's proof of the Riemann hypothesis for varieties over finite fields (Hilbert's problem 8), in: *Mathematical Developments Arising from Hilbert Problems,* Proc. Symposia in Pure Mathematics **XXVIII**, Amer. Math. Soc., Providence, RI, 1976, 275–305.

[K2] Katz, N. M., Review of [FreiK], *Bull. (N. S.) Amer. Math. Soc.* **22** (1) (1990).

[K3] Katz, N. M., L-Functions and Monodromy: Four Lectures on Weil II, *Adv. Math.* **160** (2001), 81–132.

[Lac-vF1] Laca, M., van Frankenhuijsen, M., Phase transitions on Hecke C*-algebras and class-field theory over ℚ, *J. reine angew. Math.* **595** (2006), 25–53.

[Lac-vF2] Laca, M., van Frankenhuijsen, M., Phase transitions with spontaneous symmetry breaking on Hecke C*-algebras from number fields, in: *Noncommutative Geometry and Number Theory: Where Arithmetic Meets Geometry and Physics* (K. Consani and M. Marcolli, eds.), Aspects of Mathematics E **37**, Vieweg, 2006.

[Lag] Lagarias, J. C., The Riemann Hypothesis: Arithmetic and Geometry, in [HiRoe, pp. 127–141].

[LagR] Lagarias, J. C., Rains, E., On a two-variable zeta function for number fields (Sur une fonction zêta à deux variables pour les corps de nombres), *Ann. Inst. Fourier* **53** (1) (2003), 1–68.

[LanCh] Lang, S., Cherry, W., *Topics in Nevanlinna Theory*, Lecture Notes in Mathematics **1433**, Springer-Verlag, 1990.

[Lap-vF1] Lapidus, M. L., van Frankenhuijsen, M., *Fractal Geometry and Number Theory: Complex Dimensions of Fractal Strings and Zeros of Zeta Functions)*, Birkhäuser, Boston, MA, 2000.

[Lap-vF2] Lapidus, M. L., van Frankenhuijsen, M., *Fractal Geometry, Complex Dimensions and Zeta Functions: Geometry and Spectra of Fractal Strings* (second edition), Springer Monographs in Mathematics, 2013.

[Li] Li, X.-J., A transformation of Hankel type on the field of p-adic numbers, *J. Algebra, Numb. Th. and Appl.* **12** (2009), 205–229.

[Lo] Lorenzini, D., *An Invitation to Arithmetic Geometry*, Graduate Studies in Mathematics **9**, Amer. Math. Soc., Providence, RI, 1996.

[Nau] Naumann, N., On the irreducibility of the two variable zeta-function for curves over finite fields, *C. R. Acad. Sci. Paris Sér. I Math.* **336** (2003), 289–292, and `arXiv:math.AG/0209092`, 2002.

[Ost] Ostrowski, A., Über einige Lösungen der Funktionalgleichung $\varphi(x) \cdot \varphi(y) = \varphi(xy)$, *Acta Mathematica* **41** (1) (1918), 271–284.

[Pel] Pellikaan, R., On special divisors and the two variable zeta function of algebraic curves over finite fields, in: *Arithmetic, Geometry and Coding Theory*, Proceedings of the International Conference held at CIRM, Luminy, France, 1993, 175–184.

146   *References*

[Ray] Raynaud, M., André Weil and the foundations of algebraic geometry, *Notices Amer. Math. Soc.*, September 1999, 864–867.

[Ri1] Riemann, B., Ueber die Anzahl der Primzahlen unter einer gegebenen Grösse, in [Ri2, p. 145] and translated in [Ed, p. 299].

[Ri2] Riemann, B., *Gesammelte Werke*, Teubner, Leipzig, 1892 (reprinted by Dover Books, New York, 1953).

[Roq] Roquette, P., Arithmetischer Beweis der Riemannschen Vermutung in Kongruenzfunktionenkörpern beliebigen Geschlechts, *J. reine angew. Math.* **191** (1953), 199–252.

[Ros] Rosen, M., *Number Theory in Function Fields*, Graduate Texts in Mathematics **210**, Springer-Verlag, 2002.

[Sa] Saxe, K., *Beginning Functional Analysis*, Undergraduate Texts in Mathematics, Springer-Verlag, 2002.

[fSch] Schmidt, F. K., Analytische Zahlentheorie in Körpern der Characteristik *p*, *Math. Zeitschr.* **33** (1931).

[wSch] Schmidt, W. M., *Equations over Finite Fields: An Elementary Approach*, Lecture Notes in Mathematics **536**, Springer-Verlag, 1976.

[Se] Serre, J.-P., *Local Fields,* Graduate Texts in Mathematics **67**, Springer-Verlag, Berlin, 1979.

[Sm] Smirnov, A. L., Hurwitz inequalities for number fields, *Algebra i Analiz* **4**(2) (1992), 186–209.

[So] Soulé, C., Les variétés sur le corps à un élément, arXiv:math/0304444 [math.AG], 2003.

[Ste] Stepanov, S. A., On the number of points of a hyperelliptic curve over a finite prime field, *Izv. Akad. Nauk SSSR, Ser. Math.* **33** (1969), 1103–1114.

[Sti] Stichtenoth, H., *Algebraic Function Fields and Codes*, Springer-Verlag, Berlin, 1993.

[Ta] Tate, J., Fourier analysis in number fields and Hecke's zeta-functions, in: *Algebraic Number Theory* (J.W.S. Cassels, A. Fröhlich, eds.), Academic Press, New York, 1967, 305–347.

[Tr] Tretkoff, P., Noncommutative Geometry and Number Theory, in [HiRoe, pp. 143–189].

[vF1] van Frankenhuijsen, M., Arithmetic Progressions of Zeros of the Riemann Zeta Function, *J. Number Theory* **115** (2005), 360–370.

[vF2] van Frankenhuijsen, M., The zeta function of a function field, preprint, 2007.

[V] Villa Salvador, G. D., *Topics in the Theory of Algebraic Function Fields*, Birkhäuser, Boston, 2006.

[W1] Weil, A., Sur les fonctions algébriques à corps de constantes fini, *C. R. Acad. Sci. Paris* **210** (1940), 592–594; reprinted in [W9, I pp. 257–259].

[W2] Weil, A., On the Riemann hypothesis in function-fields, *Proc. Nat. Acad. Sci. U.S.A.* **27** (1941), 345–347; reprinted in [W9, I pp. 277–279].

[W3] Weil, A., *Letter to Artin*, July 10, 1942; reprinted in [W9, I pp. 280–298].

[W4] Weil, A., *Foundations of Algebraic Geometry*, Colloquium Publications **29**, Amer. Math. Soc., New York 1946; second edition Providence, RI, 1962.

[W5] Weil, A., Number of solutions of equations in finite fields, *Bull. Amer. Math. Soc.* **55** (1949), 497–508; reprinted in [W9, I pp. 399–410].

[W6] Weil, A., Sur les "formules explicites" de la théorie des nombres premiers, *Comm. Sém. Math. Lund, Université de Lund, Tome supplémentaire* (dédié à Marcel Riesz), (1952), 252–265; reprinted in [W9, II pp. 48–61].

[W7] Weil, A., *Courbes Algébriques et Variétés Abéliennes*, Hermann, Paris, 1971. (Combined in one volume *Sur les courbes algébriques et les variétés qui s'en déduisent,* Pub. Inst. Math. Strasbourg VII (1945), 1–85, and *Variétés Abéliennes et courbes algébriques,* Actualités scientifiques et industrielles **1041**, Hermann, Paris, 1948.)

[W8] Weil, A., *Basic Number Theory*, Springer Classics in Mathematics, 1995.

[W9] Weil, A., *André Weil: Oeuvres Scientifiques* (Collected Papers) **I**, **II**, and **III**, 2nd edition with corrected printing, Springer-Verlag, Berlin and New York, 1980.

Index of notation

$.^a$ (algebraic closure), 11
f (average value), 109, 116, 122
$\langle a, b \rangle_*$ (scalar product in $\mathcal{Z}$), 108
$\| \cdot \|$ (norm on a vector space), 13
$| \cdot |_v$ (norm on a field), 12
$*$ (convolution product), 70, 109

$\mathbb{A}^*/K^*$ (idele class group), 6, 49
$\mathbb{A}, \mathbb{A}^*$ (adeles, ideles), 43

$B_r(x)$ (ball of radius r around x), 13

$\mathcal{C}$ (algebraic curve), 2, 45, 79
$c_1, \ldots, c_h$ (representatives of Cl(0)), 49
Cl(n) (idele classes of degree n), 48
$C_{\mathcal{F}\theta}$ (ultraviolet cutoff), 117, 128

Δ (diagonal), 92, 93
$D(a)$ (dilation), 127
D_ϵ (canonical pairing), 11
deg v (degree of a valuation), 31, 79–80
δ_P (extension of Kronecker delta), 111
$\mathfrak{d}_{w/v}$ (local different), 18
$d(w/v)$ (exponent of different), 18
$dx, d_v x, d_S x$ (additive Haar measure), 36, 43, 60
d^*a, d_v^*a (multiplicative Haar measure), 39, 48

E (average restriction), 50, 51
$e(w/v)$ (order of ramification), 17

ϕ, ϕ_q (Frobenius), 6, 81–83, 88, 92, 93, 98
$\mathcal{F}^*$ (Fourier transform on $\mathbb{A}^*$), 128
$\mathcal{F}, \mathcal{F}_v, \mathcal{F}_S$ (Fourier transform), 37, 44, 60
$\mathbb{F}_q$ (field of constants), 2, 31, 79
$\mathbb{F}_{q^n}$ (extension of $\mathbb{F}_q$), 2, 54
$f(w/v)$ (degree of inertia), 17

g (genus of $\mathcal{C}$), 2, 45
Gal(L/K) (Galois group), 90
$\mathcal{H}$ (Hilbert space), 109
h (class number of $\mathcal{C}$), 49

$\mathcal{I}_*, \mathcal{I}$ (inversion), 109, 110

$\mathcal{K}$ (canonical divisor), 45
K^+, K^* (additive, multiplicative group), 32, 38
K^a (algebraic closure), 11
κ_+^* ($\int f(a)\, d_v^* a = \kappa_+^* \int f(x)\, d_v x$), 39
ker $| \cdot |$ (ideles of degree zero), 49
$K(v)$ (residue class field), 17
K_v (completion), 24
k_v (canonical exponent), 36

$\Lambda_{\mathcal{C}}(s)$ (shifted zeta function), 56, **57**
$L_{\mathcal{C}}(X)$ (numerator of $\zeta_{\mathcal{C}}(s)$), 58
$L_{\mathcal{C}}(X, Y)$ (numerator of $\zeta_{\mathcal{C}}(s, t)$), 64
l_n (coefficients of $L_{\mathcal{C}}(X)$), 58
$l_n(Y)$ (coefficients of $L_{\mathcal{C}}(X, Y)$), 65
L^{sep} (separable closure), 21

$\mathcal{M}$ (Mellin transform), **70**, 110
M_Ω (infrared cutoff), 117, 126
$M_\omega, \widehat{M}_\theta$ (cutoff), 110, 111
m_w (local factor of m), 25
M_x (multiplication map), 9

$N_{\mathcal{C}}(n)$ (number of points on $\mathcal{C}$), 2
$N_{L/K}$ (norm from L to K), 10

$\mathfrak{o}$ (integral adeles), 43
$\mathfrak{o}^*$ (integral ideles), 43
$\omega_1, \ldots, \omega_{2g}$ (zeros of $\zeta_{\mathcal{C}}$), 2, 59
$\mathcal{O}_S$ (S-integers), 60
$\mathfrak{o}_v$ (local ring), 16, 31
$\mathfrak{o}_v^*$ (local units), 38

p (characteristic), 32, 79

149

Index

151

Printed in the United States
by Baker & Taylor Publisher Services